W9-DEY-920

Solvent-Free Polymerizations
and Processes

ACS SYMPOSIUM SERIES **713**

Solvent-Free Polymerizations and Processes

Minimization of Conventional Organic Solvents

Timothy E. Long, EDITOR
Eastman Chemical Company

Michael O. Hunt, EDITOR
Milliken Chemicals

American Chemical Society, Washington, DC

Library of Congress Cataloging-in-Publication Data

Solvent-free polymerizations and processes : minimization of conventional organic solvents / Timothy E. Long, editor, Michael O. Hunt, editor.

p.   cm.—(ACS symposium series , ISSN 0097-6156 ; 713)

"Developed from a symposium sponsored by the Division of Polymer Chemistry at the 214th National Meeting of the American Chemical Society, Las Vegas, Nevada, September 7–11, 1997."

Includes bibliographical references and index.

ISBN 0–8412–3591–0

1. Polymerization—Congresses. 2. Organic solvents—Congresses.

I. Long, Timothy E., 1969–   . II. Hunt, Michael O., 1961–   . III. American Chemical Society. Division of Polymer Chemistry. IV. American Chemical Society. Meeting (214th : 1997 : Las Vegas, Nev.) V. Series.

TP156.P6S64   1998
547'.28—dc21   98–34535
                                                                      CIP

The paper used in this publication meets the minimum requirements of American National Standard for Information Sciences—Permanence of Paper for Printed Library Materials, ANSI Z39.48–1984.

PRINTED IN THE UNITED STATES OF AMERICA

Advisory Board

ACS Symposium Series

Foreword

THE ACS SYMPOSIUM SERIES was first published in 1974 to provide a mechanism for publishing symposia quickly in book form. The purpose of the series is to publish timely, comprehensive books developed from ACS sponsored symposia based on current scientific research. Occasionally, books are developed from symposia sponsored by other organizations when the topic is of keen interest to the chemistry audience.

Before agreeing to publish a book, the proposed table of contents is reviewed for appropriate and comprehensive coverage and for interest to the audience. Some papers may be excluded in order to better focus the book; others may be added to provide comprehensiveness. When appropriate, overview or introductory chapters are added. Drafts of chapters are peer-reviewed prior to final acceptance or rejection, and manuscripts are prepared in camera-ready format.

As a rule, only original research papers and original review papers are included in the volumes. Verbatim reproductions of previously published papers are not accepted.

ACS BOOKS DEPARTMENT

Contents

Synthesis and Processing
in Supercritical Carbon Dioxide

Polymer Networks

Bulk Photopolymerization

Miscellaneous Solvent-Free
Polymerizations and Processes

Indexes

Preface

Significant international attention has been devoted to the reduction or elimination of traditional organic solvents from industrial processes. It has been demonstrated that some organic solvents have detrimental environmental effects on ecosystems. Waste reduction efforts are being catalyzed by environmental regulatory and product stewardship issues. In order to consolidate and raise awareness of the latest developments in solvent-free polymerizations and processes, the Division of Polymer Chemistry and the American Chemical Society organized an international symposium.

The Solvent-Free Polymerizations and Processes symposium was held at the Fall 1997 American Chemical Society meeting in Las Vegas, Nevada. More than 50 contributions from six countries were presented to very well attended meeting sessions. The importance of this research field is underscored by the breadth of the polymer chemistry contained in this publication. Research efforts in such diverse areas as controlled bulk free radical polymerization, polyester melt polycondensation, the use of "pre-polymers" in network formation, and the use of high pressure carbon dioxide as an environmentally benign solvent are all focused on the elimination or minimization of conventional organic solvents. The consolidation of these dissimilar topics will yield synergistic advances in waste reduction, novel chemistries, unique processes, and product enhancements.

The editors thank the symposium speakers and attendees, and especially the authors of the subsequent chapters. It was a pleasure to work with the authors due to their exciting research and enthusiastic cooperation in producing a timely publication. We also thank the peer reviewers who gave their time and technical insights to further improve the contributions to this book. The editors also thank the Division of Polymer Chemistry (POLY) and the Division of Polymeric Materials Science and Engineering (PMSE) for their organizational work on the American Chemical Society meeting. It was outstanding and beneficial to all of the meeting attendees. Finally, we personally thank our employers, Eastman Chemical Company and Milliken Chemical, for the opportunity to organize this very relevant symposium and symposium book over the course of the past year.

TIMOTHY E. LONG[1]
Polymer Research Division
Eastman Chemical Company
Kingsport, TN 37662

MICHAEL O. HUNT[2]
Milliken Chemicals
920 Milliken Road
Spartenburg, SC 29304

[1]Current address: Department of Chemistry, Virginia Polytechnic Institute and State University, Blacksburg, VA 24061.
[2]Current address: DuPont Lycra, Chestnut Run Plaza, P.O. Box 80702, Wilmington, DE 19880–0702.

Chapter 1

Solvent-Free Polymerizations and Processes: Recent Trends in the Minimization of Conventional Organic Solvents

Timothy E. Long[1] and Michael O. Hunt[2]

[1]Polymer Research Division, Eastman Chemical Company, Kingsport, TN 37662
[2]Milliken Chemicals, P.O. Box 1927, Spartanburg, SC 29304

Solvent-free polymerization processes and polymer melt processing have received significant attention during the past decade in both industrial and academic laboratories. Environmental concerns and federally mandated emission regulations have primarily driven this intense interest, and will continue to fuel scientific developments in the future. Research efforts in such diverse scientific areas such as controlled bulk free radical polymerization, polyester melt polycondensation, and the development of suitable polymerization reactors has focused on the reduction and often elimination of traditional organic solvents. This chapter will highlight the importance of polymer synthesis and processing in the absence of organic solvents, and focus on remaining technical obstacles to be addressed. In addition, the following chapters will overview current research activities in solvent-free processes and will illustrate the broad-based approach to the reduction of organic solvents in future polymer-based technologies.

The reduction or elimination of traditional organic solvents is a goal that crosses all boundaries in polymer science. The purpose of this symposium series book is to consolidate state-of the-art synthetic strategies in melt/solid state condensation polymerization, bulk free radical polymerization, synthesis and processing in high pressure carbon dioxide, gas phase polymerization, reactive melt processing, and network formation. Although a single reader will not find all of these topics relevant to their individual research interests, the combination of these dissimilar approaches to the minimization of organic solvents may yield some synergistic breakthroughs toward waste reduction. The commercial products of these solvent-free processes are used in such diverse applications as textile fibers, packaging resins, high performance films, medical devices, ink components, aerospace coatings, and microelectronic circuit board constituents. A viable scientific goal is the development of new polymer products that are prepared in the absence of conventional organic solvents and water, processed into useful forms in the molten state, and depolymerized efficiently to yield high purity starting materials.

Growing concerns regarding the environmental fate of industrial chemical waste streams have prompted significant research efforts to reduce or eliminate traditional organic solvents from industrial processes. Figure 1 depicts the decline of various types of air emissions and surface discharges during the past decade. Although outstanding strides in pollution reduction and awareness have been made in the last twenty-five years, significant opportunities remain. Recent advances in polymer science have included the development of bulk/solid state polymerization techniques, modification of existing polymers in traditional thermal processing equipment, applications of "pre-polymers" in network formation, and the exploration of environmentally benign solvents such as supercritical carbon dioxide. The added constraints imposed by using "non-optimum" reaction conditions require efficient reactor engineering, and strong multi-disciplinary efforts between research chemists and engineers to develop industrially feasible, solvent-free systems.

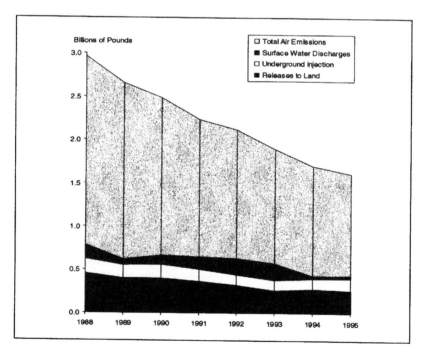

Figure 1: Distribution of Toxic Releases from 1988 to 1995 (Reproduced from Reference 1)

Due to the advent of environmental concerns and the associated regulatory and product stewardship issues, the re-examination of conventional chemical processes using organic solvents suggests that opportunities exist for improved processes. These processes not only lead to waste reduction, but also in favorable cases may lead to lower process costs through improved efficiency, higher purity products, and safer working environments. The Chemical Manufacturers Association (CMA) has advocated a formal program entitled Responsible Care that establishes business protocols to preserve the environment in a responsible manner (2). In addition, this industry-focused initiative attempts to limit the risks of the chemical business on workers and local communities, and promotes safe chemical manufacturing and transportation. Figure 2 depicts the quantities and corresponding percentages of Toxics Release Inventory (TRI) chemicals managed in 1995. The data illustrates that greater than 70% of TRI chemicals were either recycled or treated on-site. Although industrial programs such as Responsible Care have raised awareness of environmental responsibility from the management level, a fundamental scientific focus on environmentally conscious synthetic methodologies will contribute to these important initiatives. Issues including energy required to manufacture a new product, consumption of natural resources, definition of the waste stream, process or product impact on the environment, and recycling/disposal methodologies will become more important as polymer research continues into the next millennium (3).

The first section of this volume deals with the utility of melt phase and solid state polymerization strategies. The favorable economics of the elimination of a solvent as a raw material and elimination of cost incurred to dispose or recover the solvent in accordance with environmental regulations often justifies the expense of initial research efforts and subsequent process engineering work. In addition, melt phase and solid state polymerization provide a final product that does not typically

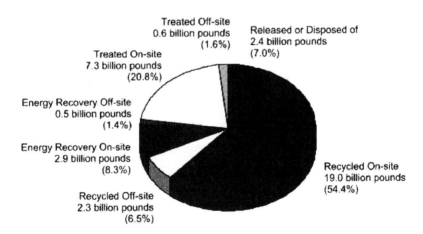

Figure 2: 1995 Toxics Release Inventory Public Data Release (1)

require further purification. The product is in a suitable form for the final consumer, and deleterious impurities derived from organic solvents are not present. For example, polymerization of polymeric amide esters at lower temperatures in the solid state reduces the propensity for deleterious side reactions and associated color body formation. In fact, poly(ethylene terephthalate) (PET) is prepared commercially in the absence of solvent and in the solid state. In addition, polyesters are viewed as ideal candidates for biodegradation and recycling due to the susceptibility of the polymeric ester functionality to subsequent hydrolysis or glycolysis. One can imagine the complexity of polymerization kinetics in the solid state including such complicating factors as simultaneous morphological changes, restricted diffusion of condensation products, and gradients in molecular weight. Many of these factors are explored in the contribution from General Electric dealing with the melt phase polymerization of polycarbonate in the absence of solvent. Despite these apparent complexities, sufficient macromolecular mobility exists to prepare high molecular weight materials at significantly lower temperatures and in the absence of organic solvents.

Bulk free radical polymerization and the utility of living free radical processes, e.g. TEMPO mediated polymerization, have resulted in a resurgence of interest in radical polymerization chemistry. The second section of this volume describes various synthetic avenues that employ living free radical methodologies. Although the discovery of nitroxide mediated free radical polymerization was first disclosed by industrial scientists at Xerox, the recent literature has exploded with novel variations and extensions. For example, bulk free radical polymerization in a living manner offers the potential to prepare conventional thermoplastics with controlled molecular architecture and molecular weight. Several chapters describe the utility of living free radical polymerization for the preparation of well-defined macromolecules. This technology offers the potential for well-defined macromolecules to impact commercial products in an economically feasible manner. The development of future polymerization catalysts and initiators that function efficiently in melt, solid, and gas phases will continue to receive significant attention.

Although this symposium was directed towards the elimination of low molar mass solvents and water, the utility of supercritical fluids represent an attractive alternative to many halogenated hydrocarbons. One can envision that the supercritical fluid, e.g. carbon dioxide, can be safely transferred and recycled. Two chapters in this section describe the use of supercritical fluids in polycondensation and free radical polymerization processes. In addition, the gas can be utilized to generate well-defined microcellular polymer products as described in relation to biodegradable polymers. Both homogeneous and heterogeneous processes are described in the presence of supercritical carbon dioxide. In addition, supercritical carbon dioxide is a viable polymer plasticizer that facilitates crystallization and impurity extractions.

Crosslinking and photopolymerization in the solid state are receiving increased attention, and two sections describe recent research efforts in these areas. For example, photopolymerization is a vital component of the fabrication of polymer dispersed liquid crystal displays, and other complex electronic circuitry. Pearce outlines a novel approach for the chemical crosslinking in a miscible polymer blend, and crosslinkable sites are selectively introduced to polymer chain ends for subsequent chain extension and crosslinking.

Polymer modification during melt processing is becoming increasingly important, and the final section deals with this very interdisciplinary research area. Peroxides are often used for the modification of hydrocarbon polymers via reactive extrusion. In fact, commercial processes involving the maleation of polyolefins lead to novel intermediates for polymer blending. A subsequent chapter elucidates the criteria for peroxide choice and processing conditions in reactive extrusion. In addition, the design of reactive extrusion profiles that are capable of handling higher melt viscosities will provide a unique reaction environment for polymer modification reactions in the absence of solvent. The new chemical processes that are developed to meet these solvent-free constraints will require very sensitive engineering controls due to the required dissipation of exothermic heat and reaction equilibria involved in both chain and step growth polymerization respectively. Thus, significant advances in process control will be an essential element of future efforts. An important issue in all cases is the ability to transport viscous polymeric melts, and efforts to prepare polymers that "flow like water" but have the mechanical properties of high molecular weight analogs will remain as a viable research goal in the future.

This edited volume of manuscripts dealing with various aspects of solvent free polymerizations and processes represents only a small fraction of the research efforts currently in progress in both academic and industrial laboratories. As one can imagine, significant advances in the development of environmentally benign chemical processes will require strong interdisciplinary programs and synergistic combinations of industrial and academic laboratories. This volume is intended to stimulate the scientific creativity of the reader through a broader understanding of some current directions in polymer science and engineering.

References

1. Chapter 5: Year-to-Year Comparisons. *1995 Toxics Release Inventory Public Data Release*; EPA 745-R-97-005; United States Environmental Protection Agency: Washington, D.C., 1997; 119
2. *Chemical and Engineering News*, May 11, 1998, p.13.

3. G. G. Bond, *Environmental Protection*, February 1998, pp. 29-34.

MELT CONDENSATION POLYMERIZATION

Chapter 2

Synthesis of Liquid Crystalline Multiblock Copolymers

Doris Pospiech[1], Liane Häussler[1], Brigitte Voit[1], Frank Böhme[1],
and Hans R. Kricheldorf[2]

[1]Institute of Polymer Research Dresden, Hohe Strasse 6, D-01069 Dresden, Germany
[2]Institute of Technical and Macromolecular Chemistry, University of Hamburg,
Bundestrasse 45, D-20146 Hamburg, Germany

The synthesis of multiblock copolymers (segmented copolymers) containing liquid crystalline segments by a solvent-free transesterification polycondensation in the melt is discussed. New, interesting materials with unique property profiles are accessible by this simple, technically relevant process.

Nowadays, a variety of special polymer architectures is known leading to different property profiles that are necessary to meet the demands of numerous applications (*1*). However, the search for new polymers is restricted by economic pressure. Therefore, blending of two or more polymers in the melt was exploited to create new materials showing a combination of properties of the blend partners. Although this method led to a variety of new commercial polymeric materials, its general drawback is the thermodynamic immiscibility of most polymers (*2*) resulting in a dramatic decrease in mechanical properties of the blends compared to the parent homopolymers because of poor interfacial adhesion between the phases and weak interfacial strength (*3*). Extensive research has been performed to improve the miscibility in polymer blends, e.g. reinforcement of the interface and decrease in interfacial tension either by adding block copolymers directly (*4-7*) or by generating block copolymers *in situ* at the interface by reactions between the blend partners (*7,8*), as well as by introducing specific interactions such as H-bonds or ionic bonds (*9*).

Synthesis of block copolymers - that is, creating new materials by chemical coupling of pre-formed oligomeric segments - is considered an intermediate way between the development of new polymers and polymer blends. Basic structures of block copolymers are shown in Scheme 1.

Diblock and triblock copolymers are usually synthesized by a sequential anionic or cationic polymerization of styrene that generates the hard blocks of the block copolymers followed by polymerization of a monomer generating the soft blocks (e.g. isoprene, butadiene etc.) (*10,11*). Linking these thermodynamically immiscible segments results in a microphase-separated morphology. Two competing effects - phase segregation driven by chemical incompatibility on the one hand and chemical linkage between the blocks on the other hand - lead to the occurrence of periodic microstructures depending strongly on the chemical nature and the length of the segments coupled (12-18). Linking liquid crystalline monomers to the poly(isoprene) block gave new poly(styrene-*co*-isoprene) diblock copolymers showing LC behavior (19), as reviewed recently (20).

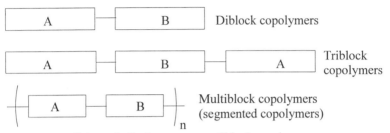

Scheme 1: Basic structures of block copolymers

If the blocks A and B are connected to form a multiblock copolymer $(AB)_n$, the morphological situation changes dramatically. Using poly(styrene-*b*-isoprene)$_n$ multiblock copolymers with constant A and B segment molecular weight, Spontak and coworkers (21,22) showed that the periodicity of block copolymer lamellae decreases with increasing n. Simultaneously, the degree of phase mixing increases. Apparently, the thermodynamic driving force for phase segregation is reduced by the higher connectivity in multiblock copolymers. On the other hand, it was demonstrated that the tensile properties of these multiblock copolymers are better than those of the diblock copolymers.

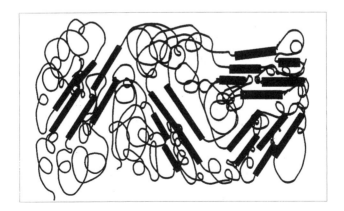

Figure 1. Structure of Multiblock Copolymers

Basically, the method of connecting segments of different flexibility in multiblock copolymers has been known since the discovery of thermoplastic elastomers (TPE).

The most important TPE are poly(urethanes) and poly(ether esters) (*23,24*), as well as poly(siloxane) block copolymers (*25*). In contrast to A-B and A-B-A block copolymers, these segmented block copolymers are usually synthesized by polycondensation of bifunctional monomers or oligomers. The connection of hard blocks with soft blocks results in a microphase-separated structure, as illustrated in Figure 1. The hard segments form domains by crystallization or by H-bonding. These domains are distributed within the amorphous matrix of the soft segments (*26*). Size and number of the hard block domains as well as the ratio of hard to soft blocks determine the elastomeric properties of the material (*23-26*).

Besides their application as thermoplastic elastomers, segmented block copolymers offer the opportunity to tailor polymer properties (*27*). Smart combination of different chemical segments allows the modification, e.g., of thermal and mechanical properties, surface tension and gas permeability.

One aspect of contemporary research has been the incorporation of liquid crystalline (LC) segments into multiblock copolymers. The main intention is the combination of advantageous properties of liquid-crystalline polymers (LCP), e.g. outstanding tensile strength and elastic modulus, dimensional stability, low coefficient of expansion etc. (*28,29*), with favorable properties of other polymers to design new materials with unique property profiles. That means in terms of Figure 1 that the hard blocks are substituted by LC-domains consisting of rigid segments uniaxially oriented within the domains. These LC-domains are randomly distributed within the matrix which can be amorphous (T_g below or above room temperature), or semicrystalline.

Synthesis principles for these LC multiblock copolymers and structures investigated over the last years will be outlined in the following chapter. Transesterification polycondensation is particularly emphasized because this method is important from the economical as well as ecological point of view. The properties of LC-multiblock copolymers (LC-MBCP) containing segments of different flexibility will be discussed and, finally, possible applications of LC multiblock copolymers will be proposed.

Synthesis of LC Multiblock Copolymers

In recent years, the number of papers dealing with the synthesis of multiblock copolymers containing LC-segments has increased considerably. Most of the LC blocks are polyesters consisting of different aromatic units randomly distributed within the polyester chain. However, block copolymers with LC-polyamides (*30-33*), LC-polycarbonates (*34*) and LC-polyethers (*35-37*) are also reported. All these types are prepared by polycondensation or polyaddition.

Overview. Some basic reactions useful for the synthesis of LC multiblock copolymers are summarized in Scheme 2. Similar reactions as for preparation of aromatic polyesters, polyamides and polyethers (*38,39*) are often applied. Most of the methods described in the following have in common that a solvent has to be used to avoid side reactions. Only transesterification polycondensation according to equation (5) can be successfully carried out in the melt.

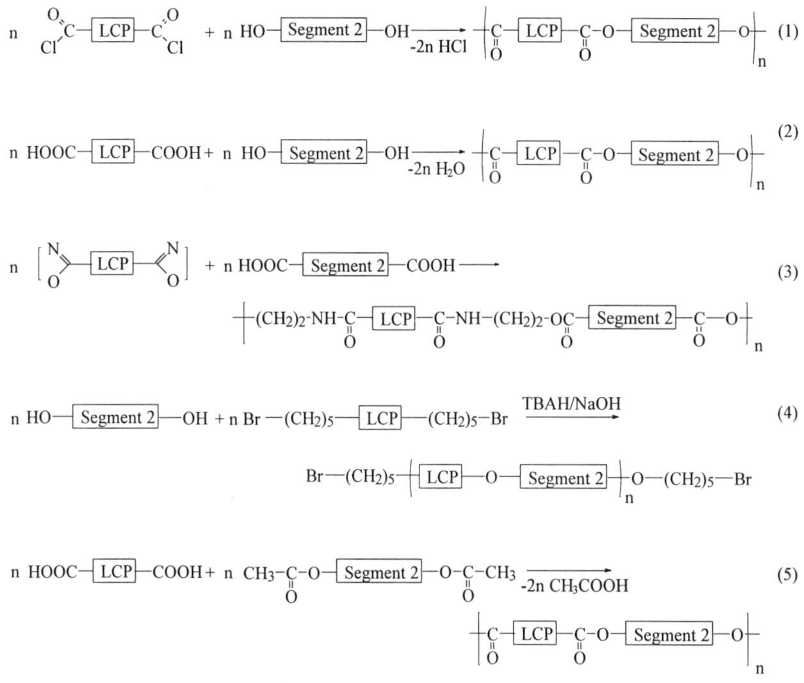

Scheme 2: Reactions Used for Synthesis of LC-Multiblock Copolymers

One of the most important polycondensation methods for LC-MBCP is shown in equation 1. The reaction of acylchloride-terminated oligomers and OH-terminated oligomers has been applied for a number of LC-MBCP, e.g., for block copolymers containing poly(p-phenylene oxide) blocks either in the main chain or as a graft in the side chain (*40-43*), poly(styrene) segments in the main chain or as a graft (*44-49*), as well as for MBCP with polysulfone and poly(phenyl-p-phenylene terephthalate) segments, (*50* and Pospiech, D., Institute of Polymer Research Dresden, unpublished results). This reaction was also successfully applied for coupling complex aromatic monomers with aliphatic polyethers (*51-53*) to give thermotropic elastomers (*51*). If the OH-terminated oligomer is replaced by aromatic diamines, MBCP with aramide sequences are formed (*32*).

The direct polycondensation of dicarboxylic acid-terminated oligomers and OH-terminated oligomers according to equation (2) often fails for preparation of MBCP because of the low reactivity of the aromatic monomers. This can be overcome by using Higashi conditions (*54*), as it was reported by Heitz et al. (*55*) for the synthesis of PPO-LC multiblock copolymers using tosyl chloride, and by

Krigbaum et al. (*30,31*) for MBCP with nematogenic polyamide segments using triphenylphosphite as a catalyst.

Bis-2-oxazolines are highly reactive with nucleophilic compounds (*56,57*) and can be used to synthesize LC-MBCP by reaction with suitable bifunctional oligomers, as illustrated in equation (3). The preparation of non-LC poly(ester amides) (*58*), poly(ether amides) and poly(thioether amides) (*59, 60*) has already been described. Segmented LC-block copolymers were obtained by reaction of oxazoline-terminated LC-pentamers with either OH-terminated polysulfone or OH-terminated poly(THF) in nitrobenzene after 40 hrs (*61* and Pospiech, D., Reichelt, N., Institute of Polymer Research Dresden, unpublished results). Polycondensation in the melt resulted in insoluble, crosslinked products by further reaction of the amide groups generated in the first reaction step.

A phase transfer catalyzed polyetherification of dielectrophiles with α,ω-bis(hydroxyphenyl)poly(ethersulfone) as described by Percec et al. (*62-64*), according to equation (4), is another method to obtain LC-segmented block copolymers.

There has been a number of efforts to apply transesterification polycondensation in the melt according to equation (5) for preparation of LC-MBCP. LC-polyester segments were formed *in situ*, i.e. by simultaneous polycondensation of acetylated diols and dicarboxylic acids with the second segment (in most cases acetoxy-terminated polysulfone) (*65-70*). The monomers forming the LC-segment were, for example, terephthalic acid, 4,4'-biphenyl diacetate, p-acetoxybenzoic acid (*65-69*), trans-cyclohexane dicarboxylic acid and chloro-hydroquinone diacetate (*70*), as well as isophthalic acid, hydroquinone diacetate and p-acetoxybenzoic acid (*71*). Because these monomer combinations give LC-polyesters with high crystallinity, transesterification polycondensation often resulted in chemically heterogeneous mixtures of block copolymer, LCP and polysulfone oligomers that could be extracted from the product. Auman et al. (*70*) detected the formation of true block copolymers with polysulfone segments of low molecular weights up to 2,300 g/mol. McGrath et al. (*67-69*) succeeded in the preparation of true segmented block copolymers by using chlorobenzene as a common solvent for both the polysulfone and the aromatic comonomers in the first reaction step. The solvent decelerates the reaction between the aromatic comonomers and consequently suppresses the formation of LCP uncoupled to polysulfone.

To avoid these complications, we used another approach to synthesize LC-MBCP. Two pre-formed oligomers have been used one of which is the LC-segment to be incorporated into the MBCP. The oligomers were chosen with respect to their ability to undergo transesterification polycondensation in the melt.

Synthesis of LC-MBCP by Transesterification Polycondensation in the Melt

Oligomer Synthesis.   Oligomers selected for synthesis of LC-MBCP by polycondensation need to have high thermal stability and stability against acetic acid generated during polycondensation. Furthermore, their melting range has to match the temperature range of the polycondensation between 210 and 300°C, and their melt viscosity has to be low enough to give good intermixing. The selected oligomeric structures are summarized in Scheme 3.

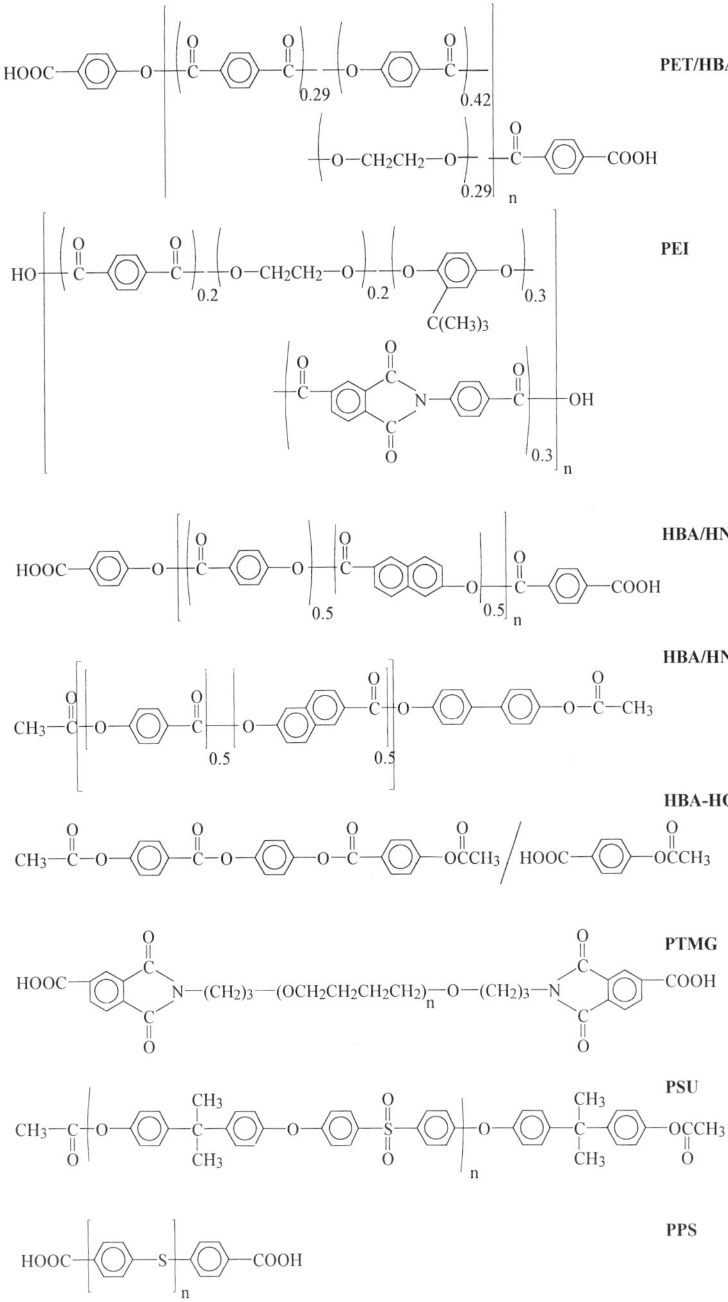

PET/HBA **(I)**

PEI **(II)**

HBA/HNA **(IIIa)**

HBA/HNA **(IIIb)**

HBA-HQ-HBA **(IV)**

PTMG **(V)**

PSU **(VI)**

PPS **(VII)**

Scheme 3: Chemical Structure of the Oligomers Used for LC-Multiblock Copolymers

Formulas (I) to (IV) illustrate the chemical structures of the LC-oligomers. (I) to (III) are copolyesters with a random sequence distribution. (I) is a semicrystalline nematic poly(ethylene terephthalate-co-oxybenzoate), according to Jackson and Kuhfuss (72). The synthesis resulted in polyesters with COOH-end groups, as discussed earlier (73), with a monomer sequence distribution near random according to ^{13}C NMR investigations (74). The molecular weight of the oligomers was controlled by adding terephthalic acid. The LC-poly(ester imide) (II) was also prepared by transesterification of PET with N-(4-carboxyphenyl)trimellitimide and tert-butylhydroquinone as recently described (Kricheldorf, H. R.; Gerken, A.; Alanko, H. *Polymer,* submitted) using an excess of trimellitimide to generate COOH-end groups. The molecular weight of poly(4-oxybenzoate-co-2,6-naphthoate)s (IIIa) and (IIIb), obtained using conditions described for the poly(HBA/HNA) synthesis (75) was controlled by adding either terephthalic acid or 4,4'-biphenyldiacetate, simultaneously resulting in COOH- or acetate end groups. Beside these oligomers with random chain segment distribution, a trimer with a regular sequence (IV) was used in combination with p-acetoxybenzoic acid for a special block copolymer (76).The LC-oligomers (I) to (IV) were chemically linked to segments (V), (VI) and (VII) with different flexibility. Poly(tetramethylene glycol) (PTMG), Formula (V), was end-capped with trimellitimide groups to suppress thermal and hydrolytic decomposition during polycondensation (76). OH-terminated polysulfone oligomers were prepared according to refs. (35,77) and acetylated as given in ref. (35), Formula (VI). COOH-terminated poly(p-phenylene sulfide) was synthesized using the procedure described by Freund and Heitz (78). All oligomers were chemically characterized by ^{13}C NMR spectroscopy, GPC and end-group titration as reported elsewhere (74,76,79,80).

Polycondensation. The oligomers discussed above were used for multiblock copolymer synthesis in combinations given in Table I.

Table I.   **LC-Multiblock Copolymers obtained by Combination of the Oligomeric Segments (I) - (VII)**

Non-LC-segment	LC-segment	Multiblock Copolymer
(V)	(IV)	PTMG-HBA-HQ-HBA MBCP (1)
(VI)	(I)	PSU-(PET/HBA) MBCP (2)
(VI)	(II)	PSU-PEI MBCP (3)
(VI)	(IIIa)	PSU-(HBA/HNA) (4)
(VII)	(IIIb)	PPS-(HBA/HNA) (5)

All multiblock copolymers under investigation were synthesized by transesterification polycondensation in the melt using comparable reaction conditions as published recently (82). The efficiency of intermixing of oligomers during melt polycondensation basically controls the conversion (degree of polycondensation). The intermixing might be disturbed by a) differences in phase behavior of the used oligomers (phase separation into isotropic and LC-oligomer phase), b) large differences in melt viscosity of the oligomers, and c) high viscosity of the reaction melt at higher degrees of conversion. In our case, polycondensation was carried out starting at 230°C under continuous nitrogen flow, followed by stepwise temperature increase to 260°C (MBCP1), 280°C (MBCP 2) or 350°C (MBCP 4 and 5), respectively. After 45 min, vacuum (133 Pa) was applied for another 90 min (60 min for MBCP 1). Though a difference of up to two orders of magnitude in melt viscosities of the starting oligomers was recorded by rheological investigation, and despite the different phase behavior of the oligomers, an uniform melt was observed by microscopic analysis of samples taken after different polycondensation times. The high melt viscosity of the reaction melt required efficient stirring to obtain high degrees of conversion.

Table II shows examples of molecular weights of different multiblock copolymer types obtained by GPC. The chemical analysis of MBCP 4 and 5 was limited by their insolubility in available solvents such as pentafluorophenol. Therefore, they are not contained in Table II.

Table II: Molecular Weights of Selected Multiblock Copolymers

MBCP	LC-segment M_w (g/mol)	2nd Segment M_w (g/mol)	Multiblock Copolymer M_n (g/mol)	M_w, (g/mol)	M_w/M_n
1a	350	1,000	38,000	55,300	2.51
1b	600	1,000	34,100	55,100	1.62
1c	350	2,000	26,400	56,300	2.13
1d	600	2,000	19,400	35,100	1.81
2a	3,200	2,100	54,000	209,000	3.87
2b	3,200	8,400	54,700	106,000	1.94
2c	5,900	5,400	64,800	108,500	1.67
2d	7,400	3,800	71,900	128,500	1.79
2 e	7,400	8,400	60,100	102,500	1.71
3a	2,900	2,560	17,100	43,300	2.53
3b	2,900	9,040	17,100	39,300	2.30
3c	7,450	2,560	13,800	33,300	2.41
3d	7,450	9,040	21,600	44,900	2.08

The molecular weights of the MBCP cannot be compared directly because they were calculated vs polystyrene. However, it is obvious that the polymers of each MBCP type have comparable molecular weight ranges. The tendencies recognizable from Table II are supported by ^{13}C NMR spectra of the MBCP. MBCP with lower molecular weight showed signs of non-reacted end groups. Segment coupling was proven by the occurrence of a new carbonyl signal of the ester bond formed during polycondensation (MBCP **1**: 149.9/168.0 ppm; MBCP **2**: 150.27/150.07 ppm; MBCP **3**: 164.78/164.68/ 165.50 ppm measured in TFA/CDCl$_3$). The molecular weights of the MBCP do not depend on the molecular weight of the segments linked, but on the type of segments used. The highest molecular weights were obtained for (PET/HBA)-PSU MBCP **2** which is due to the low melt viscosity of the liquid crystalline polyester, providing optimal reaction conditions for MBCP synthesis. The molecular weights of PEI-PSU MBCP **3** are lower because the higher melt viscosity of PEI oligomers prevents a sufficient intermixing necessary for the encounter of reactive end groups. However, there is no sign of non-reacted oligomers in the GPC curves. Side reactions were also not detected by ^{13}C NMR spectroscopy for **3**. In contrast, ^{13}C NMR spectra of the PTMG-LC-MBCP **1** having M$_w$ values in the range of 40,000 g/mol showed the presence of small amounts of aliphatic OH-end groups that were generated by chain scission of the polyether chains by COOH-end groups. This reaction disturbs the stoichiometry and consequently limits the achievable molecular weight of the polymers. However, the molecular weights of MBCP **1-3** are high enough to discuss structure-property relations.

The characterization of MBCP containing poly(4-oxybenzoate-*co*-2,6-naphthoate) segments was restricted by their insolubility in common solvents and solvent mixtures. Extraction with solvents for the oligomers (CHCl$_3$ for PSU-oligomers; pentafluorophenol or pentafluorophenol/CDCl$_3$ for HBA/HNA-oligomers; NMP for PPS-oligomers) did not indicate unconverted oligomers.

The polycondensation procedure for MBCP **2** and **5** was further simplified by using a melt kneader. It was found that kneading at 280°C (MBCP **2**), or 300°C (MBCP **5**), respectively, for 240 min, applying a weak vacuum (0.05 MPa), resulted in products with molecular weights comparable to the laboratory samples. This procedure provided the opportunity to obtain larger amounts of these interesting polymers for evaluation of their material properties.

Influence of Segment Structure and Segment Length on Properties

PTMG-LC-Multiblock Copolymers. PTMG-LC-MBCP were synthesized by polycondensation of an aromatic trimer having acetoxy-end groups and trimellitimide-terminated PTMG, as reported in ref. (*76*). The aromatic segments in these MBCP were extended by copolycondensation with p-acetoxybenzoic acid, resulting in LC-segments with a random sequence distribution. PTMG of three different degrees of polymerization were used:

$$n = \ 8 :\qquad 750 \text{ g/mol}$$
$$n = 13 :\qquad 1,450 \text{ g/mol}$$
$$\text{and}\quad n = 27 :\qquad 2,400 \text{ g/mol}.$$

WAXS and DSC investigations revealed microphase separation into an amorphous PTMG matrix and a semicrystalline, nematic phase. The glass transition of the PTMG matrix is determined by the molecular weight of the used oligomers, as illustrated in Figure 2 by DMA curves. It shifts with increasing PTMG molecular weight towards the value observed for high molecular weight PTMG. These low glass transitions provide high flexibility of the PTMG matrix, which was additionally shown by solid state NMR investigation (Holstein, P.; Pospiech, D., to be published) and by rheooptical FTIR measurements (*90*). If the PTMG molecular weight is low (DP of 8 and 13, respectively), the polymers display a thermotropic nematic mesophase above 100°C within a temperature range of 20 to 50 K. The width of the nematic region (ΔT_{ni}) is controlled by the length of the aromatic part. Copolycondensation with p-acetoxybenzoic acid forming p-oxybenzoyl sequences expands ΔT_{ni}.

The polymers behave as thermotropic elastomers. Dynamic mechanical analysis indicated the existence of a plateau region in the storage modulus curves. The storage modulus values in this region depended on the chemical composition of the MBCP. Polymers with more uniform LC-domains, formed by polymers without additional p-oxybenzoate sequences within the aromatic part, have higher storage modulus values in the plateau region (*76*).

In polymers with longer PTMG spacers (DP = 27), the PTMG is able to crystallize which was proven by WAXS and DSC. These polymers show neither liquid crystalliny in the melt nor elastomeric behavior.

It was concluded that the elastomeric behavior of PTMG-LC-multiblock copolymers is primarily caused by microphase separation into LC-domains and the amorphous PTMG matrix. The LC-domains are connected by flexible PTMG tie molecules generating a physical network which is comparable to the structure of poly(ether ester) elastomers, except that the hard blocks are not formed by H-bonds or crystallization but by uniaxially oriented LC-domains. Both the high flexibility of PTMG as well as the formation of LC-domains are requirements for the formation of LC elastomers.

Polysulfone-LC-Multiblock Copolymers.   In contrast to PTMG-MBCP the LC-domains in this case are embedded into an amorphous matrix which is rigid at room temperature (high T_g). By combined DSC, TEM and DMA investigations it was found that the glass transition temperature of the LC-segment determines significantly the degree of phase separation in the multiblock copolymers.

Poly(ethylene terephthalate-*co*-oxybenzoates) have a low T_g at 58°C and therefore, the difference to the T_g of the polysulfone oligomers is large. Consequently, the multiblock copolymers possess a high degree of phase separation. Increasing molecular weight of (PET/HBA) and PSU segments increases the phase separation further. However, phase separation takes place on a nm-scale whereas the corresponding blends show phase demixing on a µm-scale. The phase behavior of these MBCP is also determined by the molecular weight of the coupled segments. With short LC-segments, all types of phase behavior (LC, biphasic, isotropic) can be observed depending on the length of the used PSU segments. MBCP with shorter PSU segments form liquid crystalline melts, PSU segments with medium molecular weights show biphasic melts (coexistence of LC and isotropic state) and long PSU segments result in isotropic multiblock polymers. Using long (PET/HBA) oligomers

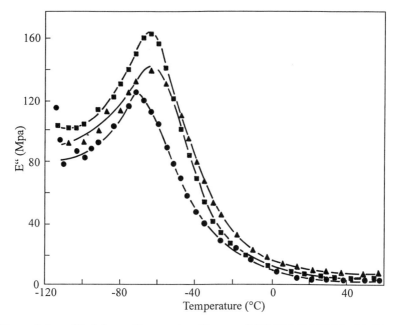

Figure 2. Loss Modulus vs Temperature Curves of PTMG-LC-MBCP with 30 mol% HBA units Obtained by Dynamic Mechanical Analysis
▲ PTMG molecular weight 750 g/mol
■ PTMG molecular weight 1,000 g/mol
● PTMG molecular weight 2,000 g/mol

results in the formation of liquid crystalline MBCP, independent of the length of the polysulfone.

In contrast to MBCP containing (PET/HBA) polyester segments, the glass transition temperature of the used LC-poly(ester imide) units is very close to those of the polysulfones. The morphology of PSU-PEI MBCP can be balanced between non-phase-separated (showing a single T_g) observed in MBCP with low molecular polysulfone segments and phase-separated by linking long segments. However, phase separation is less significant than in PSU-(PET/HBA) MBCP according to TEM, which led us conclude that the difference of the T_g of two phase-separated amorphous phases in multiblock copolymers actually reflects the degree of phase separation. The thermal behavior of PSU-(PET/HBA) and PSU-PEI MBCP is displayed for some examples in figure 3. All multiblock copolymers **3** having poly(ester imide) units show liquid crystallinity in the melt independently of the length of the incorporated polysulfone segment, indicating the strong mesogenic power of N-(4-carboxyphenylene)-4,4'-(carboxyphenoxy) phthalimide units in the LC-segments.

Both the smaller degree of phase separation as well as the LC phase behavior result in significantly higher storage modulus and T_g values of PSU-PEI compared to PSU-(PET/HBA) multiblock copolymers, as illustrated in figure 4. Whereas all investigated property parameters of PSU-(PET/HBA) MBCP range in between the levels of the parent homopolymers PSU and PET/HBA, the property level of PSU-PEI MBCB exceeds these levels, reflecting a synergistic effect of block copolymer formation.

Poly(p-phenylene sulfide)-LC-Multiblock Copolymers. In contrast to the examples of MBCP discussed in the foregoing, the non-LC-segment in PPS-MBCP is highly crystalline. PPS/(HBA/HNA) MBCP were synthesized by transesterification of acetoxy-terminated (HBA/HNA) oligomers with COOH-terminated PPS oligomers by kneading in a Brabender mixer at 300°C for 3 hrs (79). MBCP formation was indirectly concluded from the lack of non-reacted oligomers extractable by appropriate solvents. DSC investigations gave a further indication for the presence of MBCP. The thermal behavior of the block copolymers differs from that of the corresponding oligomers. Only a single glass transition was found in the MBCP occurring at higher temperature compared to PPS and Vectra oligomers. In contrast, the crystallinity of PPS did not show a significant change. However, the kinetics of PPS crystallization is influenced by the LC-segments, i.e., the crystallization is shifted to lower temperature and the rate is reduced.

A distinct phase separation into an amorphous-nematic LC-phase and a semicrystalline PPS phase in the multiblock copolymers was also proven by transmission electron microscopy (TEM). The average size of the observed spheric LCP-domains in all these MBCP is about 0.7 μm, which is about 15 times smaller than the domain size in a blend of PPS and LCP with comparable weight fractions. Compared to the MBCP that contain polysulfone units with domain sizes of some hundred nanometers, phase separation in PPS-MBCP is dramatically higher, which is presumed to result from the high tendency of PPS to crystallize.

20

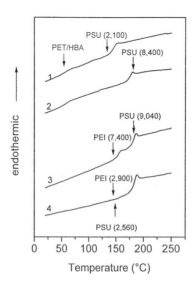

Figure 3. Glass transition Behavior of LC-MBCP under Investigation (the arrows indicate the T_g of starting oligomers)
1 PET/HBA-PSU MBCP (3,200 - 2,100)
2 PET/HBA-PSU MBCP (7,400 - 8,400)
3 PEI - PSU MBCP (7,400 - 9,040)
4 PEI - PSU MBCP (2,900 - 2,560)

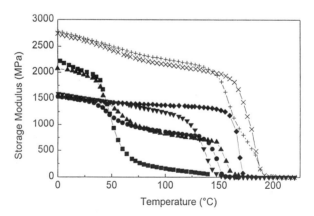

Figure 4. Influence of the Segment Molecular Weight on the Relaxation Behavior of Polysulfone - LC-Multiblock Copolymers
■ PET/HBA (7,400 g/mol)
● PET/HBA - PSU MBCP (3,200 - 3,820)
▲ PET/HBA - PSU MBCP (7,400 - 8,400)
◆ PSU (30,000 g/mol)
▼ PEI (7,450 g/mol)
+ PEI - PSU MBCP (7,400 - 9,040)
✕ PEI - PSU MBCP (2,900 - 2,560)

Application of LC Multiblock Copolymers

The LC-MBCP investigated have been examined with respect to possible applications as engineering plastics as well as compatibilizers in immiscible LCP blends. A future research prospect is the application of LC-MBCP as membrane material. Incorporation of LC-segments into the membrane polymer polysulfone (87,88) should permit control of separation process.

Engineering Materials. The property profile of LC/polysulfone multiblock copolymers as outlined in (88) particularly refers to the application as an engineering material. The polymers possess a very high thermal stability, a low thermal expansion coefficient due to the incorporation of LC-segments, and storage and Young moduli in between the range of PSU and commercial LCP. The high melt viscosity of polysulfone, having values of more than 10^4 Pa·s, is decreased by block copolymer formation with polysulfone by an order of magnitude which results in an enhanced processability of the polymer and lower energy costs. PPS-LC-block copolymers show reduced crystallinity and consequently an improved impact strength as compared to pure PPS.

Compatibilizers in Immiscible Polymer Blends. Noolandi (85) suggested some years ago that multiblock copolymers consisting of blocks chemically equal to the blended polymers would form a pancake structure at the interface between the two immisible polymers, provided that the molecular weight of the segments in the multiblock copolymer is high enough to allow the blocks to form loops into the phase-separated polymer phases. Block copolymers at the interface should decrease the interfacial tension, increase miscibility and, finally, improve the mechanical properties of such immiscible blends. The potential of LC-MBCP **2** and LC-MBCP **5** as compatibilizers in blends of the parent homopolymers, known to be immiscible, has been evaluated and is discussed in the following.

Poly(ethylene terephthalate-co-oxybenzoate)-Polysulfone Blends. The influence of (PET/HBA)-PSU MBCP **2** in blends of the parent homopolymers was elucidated using block copolymers having three different segment molecular weights. Systematic investigations of solution cast binary blends by DSC and TEM revealed, as discussed recently (Häußler, L.; Pospiech, D.; Eckstein, K.; Janke, A.; Vogel, R. *J. Appl. Polym. Sci.,* in press), that an interpenetration of the polysulfone matrix and the polysulfone phase of the block copolymer is reached (i.e., occurrence of a single T_g of polysulfone matrix and PSU phase of the multiblock copolymer), which is also observed in ternary blends of PSU/LCP and MBCP. This partial miscibility requires segment molecular weights of more than twice the entanglement molecular weight of polysulfone (i.e., MBCP **2c** of Table II), which is in accordance with earlier findings of Creton and Kramer (84) in polystyrene blends containing the corresponding diblock copolymers. Furthermore it was found that transesterification between the LC-segments of the block copolymer and the LCP blend partner obviously did not occur, pointing to only one mechanism of compatibilization.

22

Blends of high molecular weight polysulfone and high molecular weight PET/HBA with different compositions as well as different concentrations of MBCP **2e** were prepared by mixing at 260°C for 20 min. in a melt kneader. The blends were injection molded at 300°C using an Engel injection molding machine. Figure 5 shows the comparison of the tensile E-modulus (Young modulus) of blends of different LCP content without and with 5wt% MBCP. The Young moduli of all blends are lower than expected from the additivity rule which is once more a reflection of the complete immiscibility of the blends. However, the modulus values do increase upon adding MBCP. The same tendency is found in blends of PSU with the commercial polyester Vectra which means that the chemical structure of the LCP segment does not matter for compatibilization. Increasing amounts of MBCP in the blend result in further increase of the Young modulus, as illustrated in Figure 5. In the case of a PSU/Vectra blend (50/50 wt/wt), the addition of 10 % MBCP led to an increase of the modulus above the additivity curve, i.e. a synergistic effect was achieved.

Poly(4-oxybenzoate-*co*-2,6-oxynaphthoate)-Poly(p-phenylene sulfide) Blends. The influence of (HBA/HNA)-PPS MBCP in PPS/Vectra blends was investigated, too (*86*). SEM pictures of cryofractures of an injection molded PPS/Vectra blend (80/20 wt%) showed a disperse LCP phase within the PPS matrix. The particle size of the LCP particles decreased by adding an (HBA/HNA)/PPS MBCP (segment molecular weights: 2,980/1,250 g/mol), showing the emulsifying effect of the block copolymer in the immiscible blend (Figure 6). SEM furthermore indicated the presence of MBCP at the interface between PSU and LCP. The impact strength of PPS/LCP blends was increased upon adding 5 wt% MBCP. Therefore, one of the most important drawbacks of PPS (low impact strength) can be overcome by blending with LCP and the corresponding multiblock copolymer.

Summary

Transesterification polycondensation in the melt is a useful tool for synthesis of multiblock copolymers containing liquid-crystalline segments. It has been shown that these multiblock copolymers offer a great opportunity to create materials with unique property profiles. Choosing segments with suitable chemical structure as well as molecular weight permits tailoring of desired properties such as melting range and melt phase behavior (LC or isotropic), glass transition temperature, thermal stability, relaxation behavior, and, finally, mechanical characteristics of the materials.

Acknowledgments

The authors are grateful to the Saxon Ministry of Science and Art (Dresden, Germany) and Hoechst AG for financial support of the research. We would like to thank Dr. Hartmut Komber for NMR, Dieter Voigt for GPC, and Evelin Meyer for DMA investigation of the polymers. Valuable technical assistance of Kathrin Eckstein, Monika Dittrich, Kerstin Rieß, Kerstin Arnold, Christine Grunewald, Martina Franke and Hermann Laube is gratefully acknowledged.

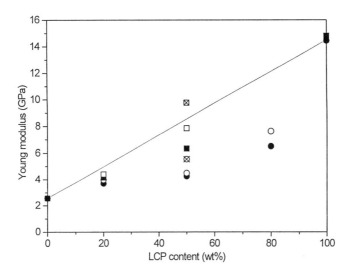

Figure 5. Young Modulus of Injection Molded Samples of LCP/PSU Blends with Multiblock Copolymer as Compatibilizer
● LCP (PET/HBA)/PSU Blends
○ LCP(PET/HBA)/PSU with 5 wt% MBCP
⊗ LCP(PET/HBA)/PSU with 10 wt% MBCP
■ LCP(Vectra)/PSU Blend
❏ LCP(Vectra)/PSU Blend with 5 wt% MBCP
⊠ LCP(Vectra)/PSU Blend with 10 wt% MBCP

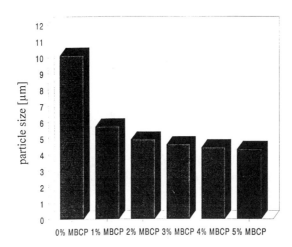

Figure 6. Particle Size Reduction in Blends PPS/Vectra (80/20) Compatibilized with a PPS-LC-Multiblock Copolymer (Molecular Weight of the PPS-Segments: 1,250 g/mol; Molecular Weight of the LC-segments: 2,980 g/mol)

24

Literature cited.

(1)   Mishra, M. K. *Macromolecular Design Concept and Practice,* Polymer Frontiers International, Inc.: New York, Vol. 1, pp 1-511.

(2)   Flory, P. J. *Principles of Polymer Chemistry,* Cornell University Press, Ithaca, New York, 1953; pp553-557.

(3)   Dai, C.-A.; Dair, B. J.; Dai, K. H.; Ober, C. K.; Kramer, E. J.; Hui, C.-Y.; Jelinski, L. W. *Phys. Rev. Lett.* **1994**, *73*, 18, 2472.

(4)   Creton, C.; Kramer, E. J.; Hadziioannou, G. *Macromolecules* **1991**, *24*, 1846-1853.

(5)   Adedeji, A.; Jamieson, A. M. *Composite Interfaces* **1995**, *3*, 1, 51-71.

(6)   Jacobson, S. H.; Gordon, D. J.; Nelson, G. V.; Balasz, A. *Adv. Mat.* **1992**,*4*,3, 198.

(7)   Xanthos, M. *Polym. Engin. Sci.* **1988**, *28*, 21, 1392.

(8)   Favis, B. D. *Canad. J. Chem. Engin.* **1991**, *69*, 61.

(9)   Eisenbach, C. D.; Hofmann, J.; Fischer, K. *Macromol. Rapid Commun.* **1994**, *15*, 117-124.

(10)   McGrath, J. E. *An Introductory Overview of Block Copolymers,* in *MMIPress Symposium Series,* **1983**, *3*, 1-16.

(11)   Canevarolo, S. V. in *Handbook of Polymer Science and Technology. Vol. 2. Performance Properties of Plastics and Elastomers, Allen, G., Berington, J. C., Eds;* Marcel Dekker, New York, 1989, pp. 3-59.

(12)   Thomas, E. L.; Lescanec, R. L. In *Self-order and Form in Polymeric Materials,* Keller, A., Warner, M., Eds., Chapman & Hall, London, 1995, pp. 147-164.

(13)   Lambert, C. A.; Radzilowski, L. H.; Thomas, E. L. *Phil. Trans. R. Soc. Lond.* **1995**, *A 13*, 1.

(14)   Bates, F. S.; Fredrickson, G. H. *Ann. Rev. Phys. Chem.* **1990**, *41*, 525.

(15)   Shull, K. R.; Kramer, E. J.; Hadziioannou, G.; Tang, W. *Macromolecules* **1990**, *23*, 4780.

(16)   Binder, K. *Adv. Polym. Sci.* **1994**, *112*, 182.

(17)   Stadler, R.; Auschra, C.; Beckmann, J.; Krappe, U.; Voigt-Martin, I.; Leibler, L. *Macromolecules* **1995**, *28*, 3080-3097.

(18)   Auschra, C.; Stadler, R. *Macromolecules* **1993**, *26*, 2171-2174.

(19)   Chen, J. T.; Thomas, E. L.; Ober, C. K.; Hwang, S. S. *Macromolecules* **1995**, *28*, 1688-1697.

(20)   Mao, G.; Ober, C. K. *Acta Polym.* **1997**, *48*, 10, 405-422.

(21)   Smith, S. D.; Spontak, R. J.; Satkowski, M. M.; Ashraf, A.; Heape, A. K.; Lin, J. S. *Polymer* **1994**, *35*, 21, 4527-4536.

(22)   Smith, S. D.; Spontak, R. J.; Satkowsi, M. M.; Ashraf, A.; Lin, J. S. *Phys. Rev. B* **1993**, *47*, 14555.

(23)   Gibson, P. E.; Vallance, M. A.; Cooper, S. L. *Morphology and Properties of Polyurethane Block Copolymers,* in *Developments in Block Copolymers-1;* Goodmann, I., Ed.; Applied Science Publ., Essex, UK, 1982; pp 217-259.

(24) Van Berkel, R. W. M.; De Graaf, S. A. G.; Huntjens, F. J.; Vrouenraets, C. M. F. *Developments in Polyether-ester and Polyester-ester Segmented Copolymers,* in *Developments in Block Copolymers-1;* Goodmann, I., Ed.; Applied Science Publ., Essex, UK, 1982; pp 217-259.

(25) Noshay, A.; McGrath, J. E. in *Block Copolymers: Overview and Critical Review,* Academic Press; New York, 1977; pp 25-29 and pp 305.

(26) Eisenbach, C. D.; Nefzger, H. *New Insights in the structure and properties of segmented polyurethane elastomers from non-hydrogen bond forming model systems,* in *Contemporary Topics in Polymer Science,* Culbertson, B. M., Ed.; Plenum Press, New York, London, 1989, Vol. 6; pp 339-361.

(27) Mülhaupt, R.; Buchholz, U.; Rösch, J.; Steinhauser, N. *Angew. Makromol. Chem.* **1994**, *223*, 47-60.

(28) Yoon, H. N.; Charbonneau, L. F.; Calundann, G. W. *Adv. Mater.* **1992**, *4*, 206.

(29) Ober, C. K.; Weiss, R. A. *Current Topics in Liquid-Crystalline Polymers,* in *Liquid Crystalline Polymers,* Weiss, R. A.; Ober, C. K., Eds.; ACS Symp. Ser. 435, Am. Chem. Soc., Washington, DC, 1990; pp 1-16.

(30) Krigbaum, W. R.; Preston, J.; Ciferri, A.; Shufan, Z. *J. Polym. Sci., Part A: Polym. Chem.* **1987**, *25*, 653-667.

(31) Jadhav, J. Y.; Krigbaum, W. R.; Ciferri, A.; Preston, J. *J. Polym. Sci.: Part C: Polymer Letters,* **1989**, *27*, 59-63.

(32) Schnablegger, H.; Kaufhold, W.; Kompf, R. T.; Pielartzig, H.; Cohen, R. E. *Acta Polymer.* **1995**, *46*, 307-311.

(33) Oishi, Y.; Nakata, S.; Kakimoto, M.-A.; Imai, Y. *J. Polym. Sci.: Part A: Polym. Chem.* **1993**, *31*, 1111-1117.

(34) Weiss, R. A.; Huh, W.; Nicolais, L. *Polym. Eng. Sci.* **1990**, *30*, 17, 1005.

(35) Percec, V.; Auman, B. C. *Makromol. Chem.* **1984**, *185*, 617.

(36) Percec, V. ; Auman, B. C. *Polym. Bull.* **1984**, *12*, 253.

(37) Shaffer, T. D.; Percec, V. *Polym. Prepr., Am. Chem. Soc. Div. Polym. Chem.* **1985**, *2*, 289.

(38) Morgan, P. W. *Condensation Polymers: Interfacial and Solution Methods,* Interscience Publ., New York-London-Sydney, 1965.

(39) *High Performance Polymers and Composites*; Encyclopedia Reprint Series , Kroschwitz, E. I., Ed., Wiley Interscience Publ., John Wiley & Sons, New York-Chichester-Brisbane-Toronto-Singapore, 1991, pp 424-431.

(40) Nießner, N.; Heitz, W. *Makromol. Chem.* **1990**, *191*, 1463-1475.

(41) Risse, W.; Heitz, W.; Freitag, D.; Bottenbruch, L. *Makromol. Chem.***1985**, *186*, 1835-1853.

(42) Fradet, A.; Heitz, W. *Makromol. Chem.* **1987**, *188*, 1613-1619.

(43) Nießner, N.; Heitz, W. *Makromol. Chem.* **1990**, *191*, 1463-1475.

(44) Heitz, T.; Höcker, H. *Makromol. Chem.* **1988**, *189*, 777-789.

(45) Sato, M.; Kobayashi, T.; Komatsu, F.; Takeno, N. *Makromol. Chem., Rapid Commun.* **1991**, *12*, 269-275.

(46) Wang, S. H., Coutino, F. M. B.; Galli, G.; Chiellini, E. *Polym. Bull.* **1995**, *34*, 531-537.

(47) Gottschalk, A.; Schmidt, H.-W., *Liquid Crystals* **1989**, *5*, 1619-1627.

26

(48)   Heitz, T.; Rohrbach, P.; Höcker, H. *Makromol. Chem.* **1989**, *190*, 3295.

(49)   Schulze, U.; Reichelt, N.; Schmidt, H.-W. *Block Copolymers with Liquid Crystalline Segments*, in *Nato Advanced Research Workshop Manipulation of Organization in Polymers using Tandem Molecular Interactions*, Il Ciocco, Italy, May 29 - June 2, 1996.

(50)   Brenda, S.; Heitz, W. *Macromol. Chem. Phys.* **1994**, *195*, 1327-1329.

(51)   Wang, J.; Lenz, R. W. *Polym. Eng. Sci.* **1991**, *31*, 111.

(52)   Mitrach, K.; Pospiech, D.; Häußler, L.; Voigt, D.; Jehnichen, D.; Rätzsch, M. *Polymer* **1993**, *34*, 16, 3469-3474.

(53)   Bilibin, A. Yu.; Piraner, O. N. *Makromol. Chem.* **1991**, *192*, 201-214.

(54)   Higashi, F.; Takahashi, I.; Akiyama, N. *J. Polym. Sci.: Polym. Chem. Ed.* **1984**, *22*, 3607-3610.

(55)   Heitz, W. *Makromol. Chem., Macromol. Symp.* **1989**, *26*, 1-8.

(56)   Wiley, R. H.; Bennett Jr., L. L. *Chem. Rev.* **1949**, *44*, 447.

(57)   Frump, J. A. *Chem. Rev.* **1971**, *71*, 483.

(58)   Kagiya, T.; Narisawa, S.; Maeda, T.; Fukui, K. *Polym. Lett.* **1966**, *4*, 257-260.

(59)   Wehrmeister, H. L. *J. Org. Chem.* **1963**, *28*, 2589.

(60)   Nishikubo, T.; Iizawa, T.; Watanabe, M. *J. Polym. Sci., Lett. Ed.* **1980**, *18*, 761-764.

(61)   Böhme, F.; Pospiech, D.; Reichelt, N. *Liquid-Crystalline Polymers with Reactive Oxazoline Groups*, 5th SPSJ International Polymer Conference (IPV 94), Osaka, Japan (Nov. 28-Dec. 2, 1994).

(62)   Shaffer, T. D.; Percec, V. *Makromol. Chem.* **1986**, *187*, 111-123.

(63)   Shaffer, T. D.; Percec, V. *Polym. Prepr., Am. Chem. Soc. Div. Polym. Chem.* **1985**, *26*, 2, 289-290.

(64)   Percec, V.; Auman, B. C.; Nava, H.; Kennedy, J. P. *J. Polym. Sci.: Part A: Polym. Chem.* **1988**, *26*, 721-741.

(65)   Lambert, J. M.; Yilgör, E.; Yilgör, I.; Wilkes, G. L.; McGrath, J. E. *Polym. Prepr., Am. Chem. Soc.* **1985**, *26*, 275-277.

(66)   Lambert, J. M.; McGrath, B. E.; Wilkes, G. L.; McGrath, J. E. *Polym. Mat. Sci. Eng.* **1986**, *54*, 1-7.

(67)   Cooper, K. L.; Chen, D. H.; Huang, H.-H.; Wilkes, G. L. *Polym. Mat. Sci. Eng.* **1989**, *60*, 322.

(68)   Cooper, K. L.; Waehamad, W.; Huang, H.; Chen, D. H.; Wilkes, G. L.; McGrath, J. E. *Polym. Prepr., Am. Chem. Soc.* **1989**, *30*, 2, 464.

(69)   Matzner, M.; Kwiatkowski, G. T.; Clendinning, R. A.; Saviar, S.; El-Hibri, M. J.; Merriam, C. N.; Cotter, R. J. *WO 9012053* (Amoco Corp.)(18 Oct., 1990), Chem. Abstr. 114(18):165139n.

(70)   Auman, B. C.; Percec, V. *Polymer* **1988**, *29*, 938-949.

(71)   Hsu, C.-S.; Chen, L.-B. *Mat. Chem. Physics* **1993**, *34*, 28-34.

(72)   Jackson, W. J.; Kuhfuss, H. F. *J. Polym. Sci., Polym. Chem. Ed.* **1976**, *14*, 2043.

(73)   Böhme, F.; Leistner; D.; Baier, A. *Angew. Makromol. Chem.* **1995**, *224*, 167.

(74)   Jancke, H.; Böhme, F.; Graßhoff, K.; Rätzsch, M.; Rafler, G. *J. Makromol. Chem.* **1989**, *190*, 3173-3183.

(75) Charbonneau, L. F.; Calundann, G. W.; Benizcewicz, B. C. *US 4,746,694.*

(76) Pospiech, D.; Komber, H.; Voigt, D.; Häußler, L.; Meyer, E.; Schauer, G.; Jehnichen, D., Böhme, F. *Macromol. Chem. Phys.* **1994**, *195*, 2633-2651.

(77) Mohanty, D. K.; Hedrick, J. L.; Gobetz, K.; Johnson, B. C.; Yilgör, I.; Yilgör, E., Yang, R.; McGrath, J. E. *Polym. Prepr., Am. Chem. Soc. Div. Polym. Chem.* **1982**, *23*, 284.

(78) Freund, L.; Heitz, W. *Makromol. Chem.* **1990**, *191*, 815-828.

(79) Kappler, D. *PhD Thesis,* Technical University of Dresden, FRG, 1997.

(80) Pospiech, D.; Häußler, L.; Komber, H.; Voigt, D.; Jehnichen, D.; Janke, A.; Baier, A.; Eckstein, K.; Böhme, F. *J. Appl. Polym. Sci.* **1996**, *62*, 1819-1833.

(81) Pospiech, D.; Häußler, L.; Eckstein, K.; Komber, H.; Voigt, D.; Jehnichen, D.; Meyer, E.; Janke, A., Kricheldorf, H. R. *Designed Monomers and Polymers,* subm. Sept. 1997.

(82) Pospiech, D.; Eckstein, K.; Komber, H.; Voigt, D.; Böhme, F., Kricheldorf, H. R. *Polym. Prepr., Am. Chem. Soc.* **1997**, *38*, 2, 398-399.

(83) Pospiech, D.; Häußler, l.; Meyer, E.; Janke, A.; Vogel, R. *J. Appl. Polym. Sci.* **1997**, *64*, 619-630.

(84) Noolandi, J.; Chen, Z. Yu. *Makromol. Chem., Theory and Simul.* **1992**, *1*, 287-389.

(85) Creton, C.; Kramer, E. J.; Hadziioannou, G. *Macromolecules* **1991**, *24*, 1846-53.

(86) Kappler, D.; Böhme, F.; Pospiech, D. *Synthesis of block copolymers of poly(p-phenylene sulfide) (PPS) and liquid crystalline poly(4-oxybenzoate-co-2,6-oxynaphthoates) and their application as compatibilizer in polymer blends, German Patent Application (DPA), 1972057.5* (Sept. 24, 1997).

(87) Lloyd, L. L. *Polyarylethers for Reverse Osmosis* in *Synthetic Membranes: Volume I Desalination,* in *ACS Symp. Ser.* 153, Turbak, A. F., Ed., Am. Chem. Soc., Washington, DC, 1981, pp. 327-350.

(88) Pospiech, D.; Häußler, L.; Meyer, E.; Janke, A.; Vogel, R. *J. Appl. Polym. Sci.* **1997**, *64*, 619-630.

(89) Miranda, N. R.; Morisato, A.; Freeman, B. D., Hopfenberg, H. B. *Polym. Prepr., Am Chem. Soc, Div. Polym. Chem.* **1991**, *32*, 3,382.

(90) Zebger, I.; Pospiech, D.; Böhme, F.; Eichhorn, K.-J.; Siesler, H. W. *Polym. Bull.* **1996**, *36*, 87-94.

Chapter 3

Polybutylene Terephthalate Modified with Diamide Segments

R. J. Gaymans and A. C. M. van Bennekom[1]

University of Twente, P.O. Box 217, 7500 AE Enschede, The Netherlands

Studied are the synthesis and thermal degradation of polybutylene terephthalate (PBT) and polybutylene terephthalate modified with diamide segments, [Benzoic acid, 4,4'-(1,4-butanediylbis(imino-carbonyl)]bis-,dimethyl ester, CAS no 1028 10-33-3. The diamide used is synthesized from butanediamine and dimethyl terephthalate in a toluene/methanol solution with Lithium methanolate as catalyst at 60-90 ºC, for 7.5 - 40 hours. The diamide has a melting temperature of 265 ºC. The polycondensation of PBT and PBT with 20 mol% of diamide units is carried out in the melt at 255 ºC under a high vacuum with Ti(OC₃H₇)₄ as catalyst. The influence of the concentration Ti-catalyst (0.02-0.28 mol%) and the diamide purity is studied. The concentration of Ti-catalyst has little effect. The presence of Lithium was found to increase the polymerization rate. Lithium thus seems to be an interesting cocatalyst. A decreasing purity of the diamide has a lowering effect on the η_{inh} and the T_m. A very effective way to obtain a high molecular weight polymer is the use of a short melt polymerization time followed with a solid state post condensation at 220º-230ºC.

The thermal stability is studied at 255º-270ºC, under nitrogen and high vacuum with a stirred melt. The degradation is followed by the change in inherent viscosity (η_{inh}), carboxylic acid endgroup concentration and amount of ester-amide interchange. The degradation constants for PBT are comparable to the literature values. The degradation constants for the PBT with diamide segments as measured by viscometry are the same as PBT but if calculated from the acid endgroup concentrations slightly higher. With degradation time the uniformity of the diamide by ester-amide interchange is lost.

Polybutylene terephthalate (PBT) is an important engineering plastic. PBT has excellent mechanical properties combined with a high crystallization rate and an ease of processing. PBT is marketed neat, impact modified, blended with other polymers as polycarbonate and glass fiber reinforced. The PBT has however a relatively low glass transition (T_g 45ºC) and low melting temperature (T_m 220ºC). Studied has been the modification of PBT with diamide segments (PBTA). The diamide segment is

---

[1]Current address: GE Plastics, P.O. Box 117, 4600 AC Bergen op Zoom, The Netherlands.

made separated before it is included in the polymer. The diamide used is synthesized from 1,4 diamino butane and dimethyl terephthalate and is called ([Benzoic acid, 4,4'-(1,4-butanediylbis(imino-carbonyl))]bis-,dimethyl   ester)   or   abbreviated T4T.dimethyl (**1**).

$$H_3CO-\overset{O}{\underset{O}{C}}-\overset{}{\bigcirc}-\overset{O}{\underset{O}{C}}-\overset{H}{N}-(CH_2)_4-\overset{H}{N}-\overset{O}{\underset{O}{C}}-\overset{}{\bigcirc}-\overset{O}{\underset{O}{C}}-OCH_3 \qquad (1)$$

T4T.dimethyl

It has been found that the T_g and T_m can be increased by modification of PBT with diamide segments (T4T.dimethyl) without losing its high crystallization rate, its ease of processing and its thermo-dimensional stability (Table I)(1-4). The rate of

Table I. Properties

	Diamide (mol%)	T_g (°C)	T_m* (°C)	T_c* (°C)	T_m-T_c (°C)	$G'_{140°C}$ (MPa)	Water** (wt%)
PBT	0	47	222	186	36	180	0.45
PBTA$_{02}$	2	51	224	191	33	-	-
PBTA$_{10}$	10	59	232	201	31	190	0.50
PBTA$_{20}$	20	70	249	219	30	250	0.85

* DSC data measured at 20°C/min heating and cooling
** Water absorption measured after 30 days, 100 RH at 25°C

crystallization of the PBTA's seems even to be higher than that of PBT. Also the blends of PBTA with PC have a higher T_m and a higher crystallization rate of the PBTA and a higher Tg of the PC phase due to lower miscibility with PBTA(5,6). The 1,4-diamino butane in the diamide has a stretched length comparable to the butanediol in the PBT. In spite of the similarity of the structure of the diamide and the ester group, the polyesteramides based on these are non isomorphous(1,4,7-9). Random copolyesteramides synthesized by co-reaction of a monomer mixture of diol, diamine and dicarboxylic acid (or esters thereof), have a low order and poor properties(7-9). Thus for PBTA with good properties it is important that the diamide segments in the polyester are uniform in length. Other uniform segments that have been incorporated in polyesters are a diimide in PBT(10), a diamide in PET(4,11), diamide in polyhexamethylene terephthalate(12) and diamides and diurethanes in aliphatic polyesters(13). A special case of polyesters with diamide segments are the alternating polyesteramides(14-18). These copolymers all have a high order and a high crystallization rate.

The faster crystallization of the PBTA (with uniform diamide segments) we expect to be due to a preordering of diamide segments in the melt (self assembling), forming very thin lamellae. These nano crystalline regions are nuclei for the crystallization of the ester segments, which follow by adjacent ordering (Figure 1)(4).

The polyesteramide with PBT and 20 mol% diamide (PBTA$_{20}$) has an average repeat length of the ester unit (x) in (**2**) of four.

$$\left[\overset{O}{\underset{}{C}}-\bigcirc-\overset{O}{\underset{}{C}}-\underset{H}{N}-(CH_2)_4-\underset{H}{N}\left[\overset{O}{\underset{}{C}}-\bigcirc-\overset{O}{\underset{}{C}}-O-(CH_2)_4-O\right]_x\right]_n \qquad (2)$$

PBTA

PBT is prepared in the melt in a two step process: transesterification of butanediol with dimethyl terephthalate (DMT) at atmospheric pressure 170°-200°C and polycondensation of the transesterified terephthalates at 250°-270°C under high vacuum. For both steps soluble titanium alcoholates are used as catalyst. The synthesis of PBTA is similar to PBT with the exception that T4T.dimethyl is added to the initial reaction mixture(1,2). T4T.dimethyl (**1**) was synthesized using diamino butane, an excess of DMT and LiOCH3 as catalyst(1,2,19). Not clear is yet what the effect of residual Lithium is on the polymerization.

The melting temperatures of the PBTA are higher than of PBT and thus the melt polymerization and the melt   processing temperatures are higher. Higher melt temperatures mean more chances of thermal degradation to take place. The limiting condition for thermal stability is the reaction time in the melt. Long reaction times in the melt can be avoided by carrying out the final part of the polymerization in the solid state. The main thermal degradation mechanisms in PBT are cyclization of butane diol and β-elimination and of which the β-elimination is at high temperatures the main mechanism(20). The β-elimination of polyesters takes place in two steps (**3**).

(3a)

(3b)

The first step involves the formation of a cyclic transition state with the formation of an acid group. In the second step butadiene is split from the unsaturated endgroup and a second acid group is formed. The second reaction proceeds much faster than the first reaction(21).

Polyamides are more thermally stable than Polyesters and from this it is expected that PBTA is as thermally stable as PBT under the condition that amide or amine groups do not interfere with the degradation reaction of the ester units. Another degradation reaction of polyesteramides is the ester-amide interchange. If ester-amide interchange in PBTA takes place, the uniformity of the diamide segment is lost and a more random copolyesteramide is obtained. For the PBTA is relevant to get insight in the ester-amide interchange rate and the effect of lost uniformity of the diamide on the polymer properties.

Studied are the synthesis of T4T.dimethyl, the synthesis of the polymers PBT and PBTA$_{20}$, the thermal degradation of PBT(A), the loss of uniformity of the diamide by ester-amide interchange and the effect of uniformity of the diamide segments on the PBTA properties.

Experimental

Materials. Dimethyl terephthalate (DMT), 1,4-butanediamine (BDA) and 1,4-butanediol (BDO) were obtained from Merck. BDA was distilled prior to use. Li(OCH3) was prepared by adding lithium to anhydrous methanol (2M). Ti(i-OC3H7)4 was distilled and diluted in anhydrous m-xylene (0.175M). N-methyl-2-pyrrolidone (NMP, Merck) was p.a. grade.

   Synthesis of T4T.dimethyl. 275 gram (1.42 mol) DMT was dissolved at 55°C in 1000 ml anhydrous toluene and 300 ml anhydrous methanol in a 2 liter flask with stirrer, condenser and nitrogen inlet. After addition of 9.77 ml (2 M) of LiOCH3 solution, 30 ml (0.30 mol) BDA in 30 ml anhydrous methanol was added dropwise. After 7.5 hours (or longer) reaction at 65°C the precipitate was separated by hot filtration and the precipitate purified by soxhlet-extraction with toluene for 6 hours. The T4T.dimethyl was recrystallized from a solution in NMP (solution of 5% at 160-180°C). The product was washed twice with acetone and subsequently dried overnight (at 60°C in vacuo).

Synthesis of PBTA$_{20}$. The reaction was carried out in a 250 ml glass flask, with N_2 inlet and mechanical stirrer. At 180°C 23.40 gram (120.7 mmol) DMT and 16.60 gram (40.23 mmol) T4T.dimethyl were dissolved in 37 ml (369 mmol) BDO during 30 min. Then 0.62 ml (0.75 M) Ti(i-OC$_3$H$_7$)$_4$ solution was added and after 30 min the temperature was raised with 15°C per 10 min to 255°C and vacuum was applied : 15 min at 15-20 mbar and 60 min at 0.1-0.4 mbar (Pirani 502 gauge). When the reaction was carried out in a 1.3 liter stainless steel autoclave (A31 type JUD25/1, Juchheim GmbH) equipped with mechanical stirrer, N_2 inlet, vacuum supply and Pirani 503 gauge, the reaction time at 180°C was prolonged to 45 min

Solid state postcondensation. The prepolymers were grinded in a Fritsch Pulverisette to a particle size of about 1 mm or smaller and dried at 100°C in vacuo overnight. Postcondensation was carried out at 20-30°C below the T_m in an oven using a low N_2 flow or reduced pressure.

Viscometry. The inherent viscosity η_{inh} was determined at a 0.5 g/dl solution in phenol/1,1,2,2-tetrachloroethane (36/64 wt%) using a capillary Ubbelohde 1b placed in a waterbath at 25.0 ±0.05°C.

DSC. A Perkin Elmer DSC7 equipped with a PE-7700 computer and TAS-7 software was used to determine the melting and crystallization transitions at a heating and cooling rate of 20°C/min The peak maximum was taken as the transition temperature T_m or T_c and the area under the curve as ΔH. For T4T.dimethyl the first heating scan ($T_{m,1}$, $\Delta H_{m,1}$) was used, for PBTA the second heating scan ($T_{m,2}$, $\Delta H_{m,2}$). Tin was used as calibration standard.

NMR. ^{1}H-NMR and proton decoupled ^{13}C-NMR spectra were recorded on a Bruker AC 250 spectrometer at 250.1 and 62.9 MHz respectively. Deuterated trifluoroacetic acid (TFA-d) was used as solvent without internal standard. ^{13}C-NMR scans (3200) were taken with an acquisition time of 2.097 sec and a 45° pulse.

Elemental analysis. The nitrogen (N) content was measured with an Elemental Analyzer 1106 (GC detection). The actual amide content (X_A) was calculated by comparison with the expected N-content, e.g. 2.549 wt% for PBTA20. The lithium (Li) content was measured by flame emission spectroscopy (AES) with a Varian AA6 or by atomic absorption spectroscopy (AAS) with a Perkin Elmer 500. Before analysis the samples were hydrolyzed under acidic conditions to disclose trapped Li.

Endgroup analysis. The endgroup content was determined with a Titroprocessor Metrohm 636 equipped with a Dosimat E635. The used titrants were trifluoromethane sulphonic acid (TFMS) in tertiary butanol (0.1M) and tetrabutyl ammonium hydroxide (TBuAOH) in isopropanol/methanol (0.1 M). For the NH_2 endgroup detection the PBTA was dissolved in phenol/water (85/15 vol%) and directly titrated with TFMS. For the COOH endgroup detection about 100 mg PBTA was dissolved in 2 ml 1,1,1,3,3,3-hexafluoro-2-propanol and diluted with 9 ml trifluoroethanol. First a certain volume of TBuAOH was added followed by a backtitration with TFMS.

Comparison of the reference with the first equivalency point (e.p.) results in the COOH content, and with the second e.p. yields the "total base" (usually NH₂).

Result and Discussion

Synthesis of T4T.dimethyl. T4T.dimethyl (1) has been synthesized in toluene/methanol (75/25-90/10 vol%) from 1,4-butanediamine (BDA) and dimethyl terephthalate (DMT)(1,2,19)(4). The formation of pure (uniform) T4T.dimethyl was promoted by the use of a large excess of DMT. T4T.dimethyl is a poorly soluble material and precipitates as soon as it is formed; further reaction is limited. The selectivity of the reaction for T4T.dimethyl is thus depended on its solubility and thus on the solvent used.

$$DMT + BDA \xrightarrow{\;LiOCH_3\;} DMT + T4 + 4T4 + T4T + T4T4 + oligomers + CH_3OH \quad (4)$$

The first precipitate (white suspension) was formed after 1 hour. On further reaction the yield increased from 65% after 7.5 hours to 85% after 40 hours total reaction time. The hot filtered mixture consisted of T4T.dimethyl, precipitated side-products (T4T4 and higher) and soluble products (DMT, T4) which had precipitated during the hot filtration step. By extraction or washing with hot toluene the soluble products like DMT-residues were removed. The impurity of the T4T.dimethyl after extraction/washing as measured by ^{13}C NMR was 4-6%, Tm = 253°-260°C, ΔHm = 110-145 J/g. Further purification by recrystallization from hot NMP resulted in an impurity content of <2 %, Tm = 263°-265°C, ΔHm = 165-173J/g. The Lithium content in the recrystallized T4T.dimethyl was 40 ppm.

The LiOCH₃ catalyst is easily inactivated by water, present in the monomers, in the reaction set-up or in the solvents, e.g. anhydrous toluene could contain 0.03 wt% water at maximum. These hydrolytic reactions (5 and 6) lead to lithium hydroxide (LiOH) and lithium terephthalate (LiTe)(22) which are less soluble and, therefore, less catalytically active.

$$Li\,OCH_3 + H_2O \rightleftharpoons LiOH + CH_3OH \quad (5)$$

The catalyst concentration used was 5 mol%/mol diamine. Probably the catalyst concentration can be reduced if less water is present.

Purity of T4T.dimethyl. The amount of side-products has been calculated via triad analysis of ^{1}H-NMR and ^{13}C-NMR scans. The chemical shifts of the terephthalic protons and carbons depend on the sequence of amide- or ester groups in the para-position of the ring (7).

The ^{1}H NMR is very sensitive but the aromatic proton peaks are partly overlapping[2,19]. The ^{13}C-NMR is in general less sensitive than ^{1}H-NMR. An advantage of ^{13}C-NMR is that the terephthalic carbons showed single peaks (Figure 2). Therefore this method seemed to be as sensitive as ^{1}H NMR. With ^{13}C-NMR only structurally similar carbons were compared because of possible differences in

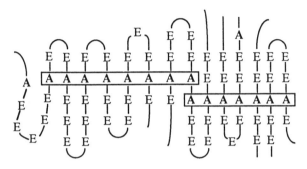

1. Crystalline ordering of amide units in a polyesteramide (Reproduced with permission from reference 4. Copyright 1997 Butterworth Heineman.)

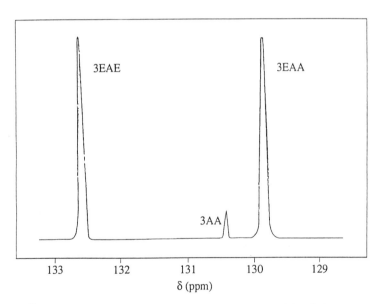

2. ^{13}C-NMR spectrum of T4T.dimethyl with longer amide oliglomers as impurity

relaxation time. The chemical shifts of the terephthalic protons and carbons are given in Table II.

Table II. Assignment of the chemical shift δ to aromatic protons and carbons of T4T.dimethyl in TFA-d (without TMS) which depended on the neighboring R group: R*-T-R

Code	R*	R	δ [ppm] in H-NMR		δ [ppm] in ^{13}C-NMR	
3^{AA}	amide	amide	8.05	singlet	130.4	singlet
3^{EAA}	amide	ester	7.98+8.02	doublet	129.7	singlet
3^{EAE}	ester	amide	8.30+8.34	doublet	132.6	singlet
3^{EE}	ester	ester	8.28	singlet	132.0	singlet

The non uniformity in the diamide (X_{AA}/X_A) has been determined via the integrated peaks of the 3-carbons (Equation 1).

$$X_{AA} / X_A = \frac{3^{AA}}{3^{AA} + 3^{EAA}} \times 100\% \ [mol\%]$$

(eq.1)

The ^{13}C-NMR scan of non-purified T4T.dimethyl (Figure 2) clearly shows the presence of terephthalic amide-amide sequences. Different side-products (4T4, T4T4, T4T4T etc.) could have contributed to the 3^{AA} peak. Endgroup analysis of this T4T.dimethyl before recrystallization affirmed the presence of amino endgroup containing side-products. Purification by recrystallization reduced the 3^{AA} peak and thus the amide block content, from about 6-4% to >2 mol%.

Thermal analysis by DSC showed for non-purified T4T.dimethyl the presence of DMT-residues at 140°C and a broad melting transition at 240°-260°C. Purification by recrystallization resulted in sharper T_m transition, at a higher temperature and with an increased heat of melting. The melting temperature as function of X_{AA}/X_A (the non uniformity in the diamide) shows a steady decrease of T_m with X_{AA}/X_A (Figure 3).

Polycondensation. The polymerization of PBTA$_{20}$ from T4T.dimethyl, BDO and DMT has been divided into three stages: transesterification, condensation to (pre)polymers in the melt, and solid state postcondensation. The results of PBTA$_{20}$ are compared with PBT.

Transesterification. At moderate temperatures (160-200°C) the methyl esters of T4T.dimethyl and DMT were transesterified with BDO into hydroxy butyl esters. The interchange reaction was catalyzed by Ti(i-OC$_3$H$_7$)$_4$. Higher temperatures would lead to the loss of DMT by sublimation and the loss of BDO by cyclization into tetrahydrofuran. As T4T.dimethyl melts at 265°C it had to be dissolved in the mixture of molten DMT and BDO (1.5-2.5 times excess). PBTA$_{20}$ with 20 mol% amide was synthesized from 1 mol T4T.dimethyl, 3 mol DMT and 8 mol BDO. For the synthesis of PBTA with a higher amide content, less BDO is stoichiometrically required but for the dissolution of T4T.dimethyl a larger excess of BDO was necessary. The effect of time and temperature of the transesterification on the inherent viscosity of PBTA$_{25}$ (25 mol% amide) suggest that the optimal conditions for transesterification, 30 minutes at 180°C(1). When most of the methanol was distilled off, the temperature was raised with 1.5 °C/min to 255°C. Longer transesterification times or slower heating rates would result in solidification of oligomer or polymer.

Once solidification occurs, further reaction is strongly retarded and remelting requires a higher temperature.

Melt polycondensation. At high temperatures, preferably 250-260°C, the condensation of the oligomeric mixture into high molecular weight polymer took place. The equilibrium of this transesterification reaction was favored by the removal of the condensation product butanediol (b.p. 230°C at 1 bar). At 255°C the pressure was gradually reduced to 0.2-0.5 mbar to prevent foaming of the reaction mixture. The removal of BDO was favored by thin film-like melt layers that reduced the diffusion path through the viscous melt. Therefore, the surface to volume ratio and the surface renewal efficiency are critical parameters for the final molecular weight of the polymer.

It is known that during the melt polymerization of PBT also degradation of the polymer takes place by β-elimination with the formation of carboxylic acid groups(23). Therefore the melt polymerization time is limited. A standard melt polymerization time is 1-2 hours. An option is to shorten the melt polymerization time at high temperatures. In this way the amount of thermal degradation is reduced. In the subsequent solid state polymerization the polymerization rate is possibly relatively higher than the degradation rate.

Solid state postcondensation. In the solid state, ester-interchange as well as direct esterification and amidation of carboxylic acid endgroups with hydroxyl and amino endgroups can take place. In the solid state polymerization of polyamides, the diffusion of the endgroups is the rate-determining step(24,25). Next to endgroup diffusion, the diffusion of the condensing diol limits the reaction(26) the most. In the solid state polymerization the molecular weight of the polymer increases with the logarithm of the polymerization time(23,27).

In the solid state polymerization of PBTA the reaction temperature was kept at about 25°C below the melting temperature of the polymer in order to favor the diffusion rate of reactive endgroups and evaporation of the condensation products(28). The use of a reaction temperature nearer the melting temperature would lead to particle sticking and thus to a surface reduction. Either vacuum could be applied or a N_2 flow to lower the partial pressure of butanediol. During the postcondensation, the size and the structure of the crystallites may change (annealing effect). An increase of crystallinity reduces the mobility of the endgroups. We studied the effect of reducing the melt polymerization time from 60 min to 15 min on the inherent viscosities and the endgroup concentration after the melt polymerization and after the solid state polymerization (Table III).

Table III. Effect melt polycondensation time at 255°C on solid state post condensation

polymer	melt polycondenzation				solid state postcondenzation			
	Time	P	η_{inh}	COOH	NH_2	conditions	P	η_{inh}	COOH	NH_2
	(min)	[mbar]	[dl.g^{-1}]	[meq.g^{-1}]	[meq.g^{-1}]		[mbar]	[dl.g^{-1}]	[meq.g^{-1}]	[meq.g^{-1}]
PBTA$_{20}$	60	2.5	0.78	0.063	0.055	218°C/48 hrs	3.0	0.80	0.076	0.033
PBTA$_{20}$	15	3.0	0.58	0.010	0.035	225°C/20 hrs	0.1	1.08	0.043	0.032

The results indicate that for obtaining high molecular weight polymers a short melt polymerization time is sufficient or even better.

The prepolymer (after 15 min of high vacuum at 255°C) was found to have a considerably lower content of carboxylic acid endgroups. As a result of this it can be expected that the [OH]/[COOH] endgroup ratio is higher which is more favorable for postcondensation. Pilati(27) reported a most optimal ratio between 2 and 10. The amine endgroup concentrations are relatively low and decrease a little with solid state polycondensation.

Kinetics of the melt polymerization. In the melt polymerization of PBT, the polymerization and degradation take place simultaneously. Starting from low molecular weight prepolymers the polymerization in the initial stage is fast and at long reaction times the degradation is dominant. PBT prepolymer has as it reached the polymerization conditions a η_{inh} of 0.1. With reaction time the η_{inh} increased strongly, having a maximum after 3 hours at 255°C. After that the viscosity decreased due to then higher thermal degradation rate. The duplicate reaction is in line with the first reaction. The evolution of the inherent viscosity η_{inh} of PBT has been determined by periodically taking samples from the reaction mixture.

One series of PBTA$_{20}$ polymerization's was carried out with different experiments for each reaction time. While another series consisted of taking samples at one reaction at different times during the polymerization. The results of the two methods are very comparable. PBTA$_{20}$ appeared to have an initially stronger increase in inherent viscosity with time compared to PBT (Figure 4). There are several possible reasons for this increased reaction rate. As the T4T.dimethyl has a higher molecular weight than DMT, the polymer molecular weight with the same ester conversion is higher. With a T4T concentration of 20% this effect is a 20% increase in molecular weight, which does not explain a doubling of the rate. Further, the presence of amide groups may have increased the polarity and as a consequence the reactivity of the mixture. It is also possible that Li-residues in the T4T.dimethyl may act as a cocatalyst in the polymerization of PBTA. This cocatalyst effect is affirmed by the synthesis of PBT with 160 ppm Li (+ in Figure 4). From the literature it is known that sodium is a cocatalyst in the PBT polymerization(29). With Lithium not only is the reaction rate higher but also the maximum attainable molecular weight is higher. This suggests that Lithium has only an effect on the polymerization rate and not on the degradation rate.

Figure 4 also shows that PBTA$_{20}$ attained a lower maximum viscosity than PBT. The lower maximum molecular weight of PBTA$_{20}$ must be due to an imbalance in reactive groups or a faster degradation reaction. The decrease with reaction for long polymerization times is for PBTA nearly the same as for PBT so the degradation with chain cleavage does not seem to be the cause. The Lithium is not the cause either as with Li in PBT high molecular weights are attained. Possibly in PBTA some of the butanediamine is not present as a diamide but as the cyclic mono functional pyrrolidine group.

Effect of the T4T purity. As already mentioned, the synthesis of T4T.dimethyl required a relatively large amount of LiOCH$_3$. The amounts of lithium-residues and amino-endgroup containing and/or oligomeric side-products were reduced by purification of T4T.dimethyl. In Table IV it is shown that the recrystallization step had a positive effect on the final inherent viscosity and melting temperature of PBTA$_{20}$.

The amino endgroup containing side-products in T4T.dimethyl were suspected to have an effect on the molecular weight of PBTA. A number of processes could have occurred, such as inactivation of titanium catalyst, aminolysis of ester bonds, reaction with carboxylic acid endgroups or cyclization into chain terminating pyrrolidine-endgroups. Inactivation of the titanium catalyst by NH$_2$ endgroups is not likely as the

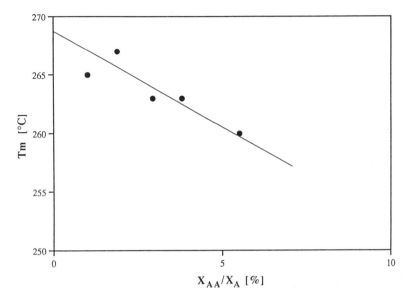

3. Melting temperature of T4T.dimethyl as function of longer amide oliglomer concentration.

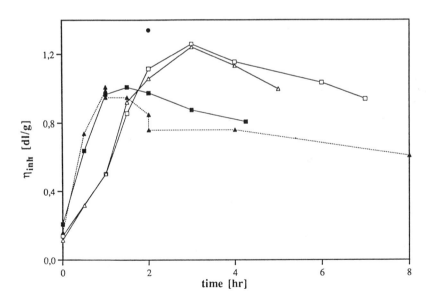

4. Evolution of inherent viscosity of polycondensation reactions at 255°C as function of reaction time, samples were taken during the reaction: $\triangle$, $\square$, PBT; $\bullet$, PBT+ 160 ppm Lithium; $\blacktriangle$, $\blacksquare$, $PBTA_{20}$

polymerization rates of PBTA are higher than for PBT. The reaction of amine endgroups with acids or esters is not expected to decrease the molecular weight.

Table IV. Comparison of $PBTA_{20}$ synthesized from non-purified T4T.dimethyl and recrystallized T4T.dimethyl

| polymer | recrys. | T4T.dimethyl | | | | | PBTA$_{20}$ | | | |
		Li [ppm]	NH$_2$ [meq/g]	X$_{AA}$/X$_A$ [%]	T$_{m,1}$ [°C]	η_{inh} [dl/g]	NH$_2$ [meq/g]	COOH [meq/g]	X$_{AA}$/X$_A$ [%]	T$_{m,2}$ [°C]
PBTA$_{20}$	no	99	0.138	6.1	256	0.70	0.039	-	7.5	241
PBTA$_{20}$	yes	42	0.040	1.9	267	0.89	0.029	0.043	5.1	251

Possibly some diaminobutane is present as cyclic monofunctional pyrollidine as is the case in the synthesis of PA4,6(30).

Effect of titanate catalyst on the polycondensation. The transesterification of PBT and PBTA will hardly take place without catalyst. Otton(7) reported that as little as 1 ppm Ti as Ti(OR)$_4$ should be effective for the synthesis of PBT. The optimum Ti-concentrations in the PBTA20 polymerization seem to be 0.068 mol% Ti(i-OC$_3$H$_7$)$_4$ per diester (DMT and T4T.dimethyl) which resulted in about 120 ppm Ti in the polymer (Figure 5).

Degradation

Viscometry. In Figure 4 the η_{inh} are given for the melt polymerization at 255°C. Next to that are studied the degradation in the melt of polymers synthesized for 1 hour at 255°C. The degradation of these polymers was carried out at 270°C under vacuum or in a nitrogen atmosphere (Figures 6 and 7). For each data set a different polymer and for each data point a different experiment was carried out. A low and a high molecular weight PBT was used for the degradation studies. PBT$_{255vac}$ (255°C/vacuum) and PBT$_{270vac}$ (270°C/vacuum) had maximum η_{inh}'s after 2 hours. The viscosities for PBT$_{270vac}$ were considerably lower than PBT$_{270.N2}$. This suggest that the PBT$_{270vac}$ with the low starting molecular weight had an imbalance in endgroups. The PBTA$_{20}$ degradation studies at 270°C and the same starting molecular weight show little difference between degradation in vacuum compared to in N$_2$ (Figure 7). The attained maximum viscosities for PBTA are lower than for PBT. The viscosity changes suggest the simultaneous occurrence of condensation and degradation reactions in the experiments.

COOH and NH$_2$ endgroups. The formation of carboxylic acid (COOH) endgroups during prolonged reaction in the melt is shown in Figures 8 and 9. From the increase of COOH endgroup concentration with time the degradation rate of PBT in vacuo or N$_2$ atmosphere was determined (Figure 8). The degradation rates at 270°C in vacuum and N$_2$ were comparable and higher than at 255°C. The PBT$_{270vac}$ had The higher COOH concentration PBT$_{270vac}$ suggest that this lower molecular weight sample had an imbalance in endgroups. In PBTA$_{20}$ the formation of COOH endgroups increased linearly with time, similar to PBT (Figure 9). The rate of COOH formation at 270°C in vacuum was higher than at 255°C and the rate for 270°C in nitrogen is in-between.

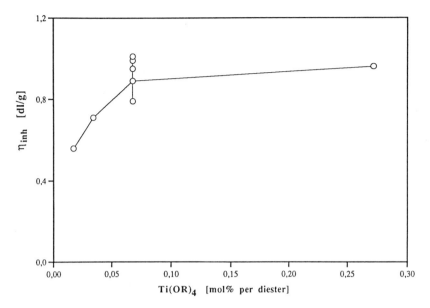

5. Effect of Ti(OR)4 concentration on the inherent viscosity of PBTA$_{20}$ (1 hr at 255°C/vacuum) (Reproduced with permission from reference 2)

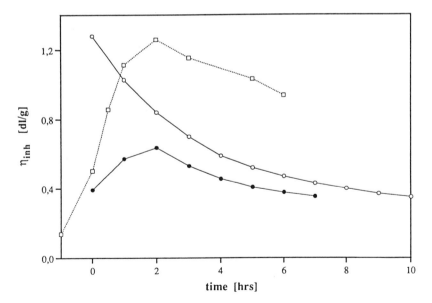

6. The inherent viscosity of PBT during prolonged heating: ☐, 255°C/vacuum; ○, 270°/N$_2$; ●, 270°/vacuum (Reproduced with permission from reference 3)

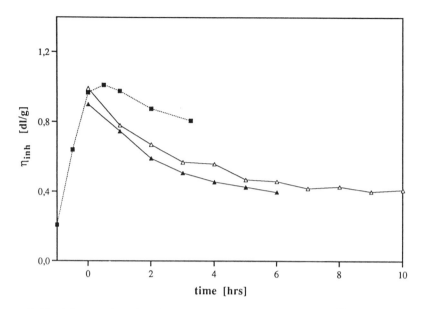

7. The inherent viscosity of PBTA$_{20}$ during prolonged heating: ■, 255°C/vacuum; △, 270°/N$_2$; ▲, 270°/vacuum (Reproduced with permission from reference 3)

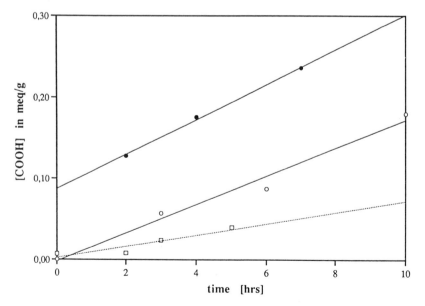

8. The carboxylic acid endgroup concentration of PBT versus time: : □,255°C/vacuum; O, 270°/N$_2$; ●, 270°/vacuum (Reproduced with permission from reference 3)

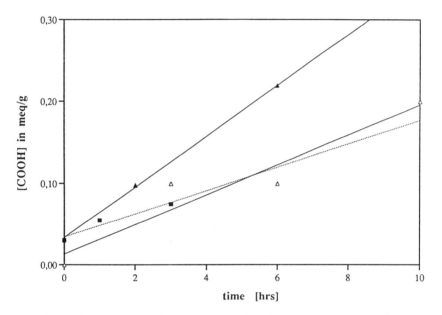

9. The carboxylic acid endgroup concentration of PBTA$_{20}$ versus time: ■, 255°C/vacuum; Δ, 270°/N$_2$; ▲, 270°/vacuum (Reproduced with permission from reference 3)

The concentration of NH_2 endgroups in $PBTA_{20}$ is very low and decreased a little with reaction (Figure 10). The degradation condition had little effect. The number of NH_2 endgroups may have been reduced by its amidation with COOH endgroups or by its cyclization into pyrrolidine endgroups. It can not be excluded that new NH_2 endgroups are formed via hydrolysis or alcoholysis of amides. But if these hydrolysis reactions take place they occur at a low rate. The pyrrolidine endgroup concentration has as yet not been determined.

Degradation kinetics. The main mechanism of degradation of PBT is by β-elimination (3). The degradation is in two steps. The second step is much faster than the first step ($k_2 \gg k_1$) and thus the first step is the rate determining step(9). The rate could be described by a first order reaction(11), Equation 2.

$$\frac{1}{\overline{X_n}(t)} = \frac{1}{\overline{X_n}(o)} + k_1 t \qquad \text{(eq.2)}$$

The degradation rate constant (of the β-elimination) could be determined from the change in viscosity and from the change in COOH endgroup content. For each cleavage of PBT by β-elimination two acid groups are formed. We assumed that the degradation mechanism for PBT is also applicable to PBTA.

Viscometry. The number-average degree of polymerization $\overline{X_n}$ (or $\overline{M_n} / M_o$) of PBT and $PBTA_{20}$ has been calculated from the inherent viscosity measured in phenol/1,1,2,2-tetrachloroethane (36/64 wt%) at 25°C. We used a Mark-Houwink relation (Equation 3), which was valid for phenol/1,1,2,2-tetrachloroethane (60/40 wt%) at 25°C(31).

$$[\eta] = 2.15 \times 10^{-4} \overline{M_n}^{0.82} \qquad \text{(eq.3)}$$

The intrinsic viscosity [η] was calculated from the inherent viscosity η_{inh} with a one-point-method of Raju(32) (Equation 4). The concentration c was 0.5 g/dl and the relative viscosity was calculated from: $\ln \eta_{rel} = c \times \eta_{inh}$. When $\eta_{rel} < 1.3$, K is 0.14 and when $1.3 < \eta_{rel} < 1.8$ K, is 0.12.

$$\log[\eta] = \log\left(\frac{\eta_{rel} - 1}{c}\right) - K(\eta_{rel} - 1) \qquad \text{(eq.4)}$$

The difference in [η] determined by extrapolation of η_{inh} to zero concentration with [η] via Raju remained within 3%. $\overline{X_n}(0)$ has been taken at the time where the maximum value of viscosity was attained.

In Figure 11 the reciprocal $\overline{X_n}$ derived from the viscosity data is presented as a function of the degradation time at 255°C or 270°C. Initially the molecular weight increased; at longer reaction times degradation is the main mechanism. From the slopes of these curves the degradation rates can be calculated (Table IV).

COOH Endgroups. By each cleavage of PBT by β-elimination two acid groups are formed. With this method for the calculation of $d\overline{X_n}/dt$ Equation 5 taken

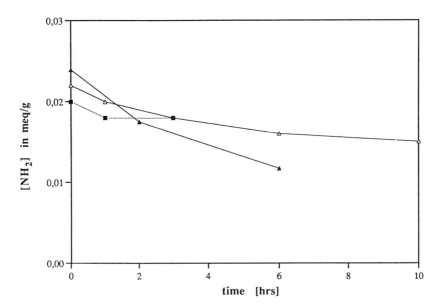

10. The amine endgroup concentration of PBTA$_{20}$ versus time: ■, 255°C/vacuum; Δ, 270°/N$_2$; ▲, 270°/vacuum (Reproduced with permission from reference 3)

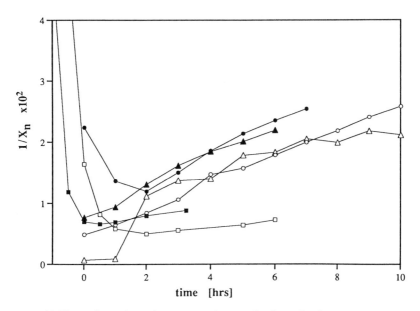

11. The reciprocal number average degree of polymerization:
: □, PBT, 255°C/vacuum; O, PBT,270°/N$_2$; ●, PBT, 270°/vacuum
: ■, PBT A$_{20}$, 255°C/vacuum; Δ, PBT A$_{20}$, 270°/N$_2$; ▲, PBT A$_{20}$,
270°/vacuum (Reproduced with permission from reference 3)

$$\frac{dX_n}{dt} = \frac{d\frac{2}{(COOH)}}{dt} \qquad \text{(eq. 5)}$$

One can also rewrite Equation 2 with Equation 5 to

$$[COOH]_t = [COOH]_o + 2k_1 t \qquad \text{(eq.6)}$$

The linearity was verified for our degradation experiments of PBT (Figure 8) and PBTA$_{20}$ (Figure 9). The results indicate that the degradation reaction is first order and that the β-elimination in PBT and PBTA20 is not catalyzed by the generated COOH endgroups.

Degradation constants. The degradation rate constant k_1 from viscosity measurements (Figure 11) and k_1 from COOH endgroups measurements (Figures 8 and 9) were calculated from the slopes via a least-squares linearization. In Table V the results for PBT and PBTA$_{20}$ are given as well as data from the literature for PBT, PHT (polyhexamethylene terephthalate) and the alternating polyesteramide 6NT6. Our degradation rate constants for PBT are of the same order of magnitude as the values reported by Passalacqua(21) and Deveaux(31), although their degradation experiments were carried out in a different way. Passalacqua had determined the degradation rate of PBT in a glass ampoule placed in an oven with an accurate temperature control ($\pm 1°C$). Deveaux, on the other hand, used a capillary rheometer at $550\ s^{-1}$ with presumably good heat transfer in the barrel. Our slightly lower degradation constants might be due to the difference in set-up and/or different catalyst concentration.

Table V. Degradation rate constants k_1 of PBT and PBTA$_{20}$ compared to PBT, PHT and 6NT6, determined by viscometry and endgroup analysis respectively

Polymer	T [°C]	conditions	k_1 [$10^{-7}\ s^{-1}$] from visco.	k_1 [$10^{-7}\ s^{-1}$] from COOH--	Reference
PBT	270	N$_2$/stirred	6.0	5.2	--
PBT	270	vacuum/stirred	7.6	6.4	--
PBTA20	255	vacuum/stirred	2.3	4.6	--
PBTA20	270	N$_2$/stirred	5.1	5.5	--
PBTA20	270	vacuum/stirred	6.9	9.5	--
PBT	260	vacuum/sealed	8.4	8.4	Passalacqua(21)
PBT	280	vacuum/sealed	37.1	34.5	Passalacqua(21)
PBT	253	N$_2$/rheometer	9.1	-	Deveaux(31)
PBT	267	N$_2$/rheometer	26.8	-	Deveaux(31)
PHT*	260	N$_2$/stirred	6.0	-	Rafler(33)
PHT*	280	N$_2$/stirred	17.1	16.9	Rafler(33)
6NT6**	260	vacuum/sealed	21.0	-	Pilati(34)
6NT6**	280	vacuum/sealed	86.1	-	Pilati(34)

*) polyhexamethylene terephthalate
**) alternating polyesteramide based on PHT

The degradation constants for $PBTA_{20}$ as measured with viscometry are the same as for PBT. The degradation constants of PBTA as determined from COOH end groups were somewhat higher than for PBT with COOH and PBTA with viscometry. This increase is much less than the factor 3-5 increase observed for 6NT6 compared to PHT (Table IV)(21). Pilati(34) suggested that amide groups in a terephthalic para-position to ester groups destabilized the ester bond.

Ester-amide interchange. At high temperatures ester-amide interchange can take place with as consequence a loss in uniformity of the amide segment length due to the formation of oligomers. With ^{13}C-NMR the amide block fraction (degree of disorder) X_{AA}/X_A in the polymer was determined as function of degradation time (Figure 12). With increasing time and temperature, the X_{AA}/X_A concentration increased. The atmosphere in the reactor (nitrogen flow or reduced pressure) had little effect on the rate of interchange.

The catalytic effect of the Titanate compound on the ester-amide interchange was tested by increasing the $Ti(OC_3H_7)_4$ in $PBTA_{20}$ from 120 ppm to 500 ppm Ti. The X_{AA}/X_A for the 500 ppm Ti polymer after 1 hr at 255°C was 4.6% that is comparable to the values of 5 - 6% for the 120 ppm Ti polymers. The larger amount of Titanates did not accelerate the ester-amide interchange reaction.

The melting temperature of starting material T4T.dimethyl decreases with X_{AA}/X_A (Figure 3). The melting temperature of $PBTA_{20}$ decreases strongly too with the amide non uniformity X_{AA}/X_A (Figure 13). Probably, the longer amide blocks in $PBTA_{20}$ have a disturbing effect on the nano ordering of the diamides.

Conclusions

PBT and PBT modified with diamide segments have been prepared by polycondensation. The diamide monomer was prepared with excess DMT in a toluene/methanol mixture with lithium methanolate as catalyst. The precipitation of the diamide from solution by being insoluble, is a selective method of stopping the reaction. The uniformity of T4T.dimethyl after synthesis was about 95%. The amount of side products (probably T4 and T4T4) and lithium catalyst residue was further reduced by recrystallization from NMP. Purified T4T.dimethyl had sharper melting peaks located at higher temperatures.

The purity of T4T.dimethyl was important for the uniformity of amide-block length in PBTA. The increase in uniformity of the amide segments led to a higher melting temperature of PBTA.

The polymerization of PBT and $PBTA_{20}$ were carried out at 255°C and high vacuum with $Ti(OC_3H_7)_4$ as catalyst. Optionally a solid state post condensation was carried out at 220°-230°C. High molecular weights could be obtained with melt polymerization and a combination of melt polymerization and solid state polymerization. For PBTA particularly interesting seem to be the combination of a short melt polymerization time followed with a solid state polymerization.

The polymerization of PBTA as compared to PBT had a higher polymerization rate but a lower maximum viscosity. It seems that the higher polymerization rate is due to the cocatalytic effect of Lithium still present in the T4T.dimethyl.

The thermal stability of PBT and PBTA has been followed during prolonged heating of the melt in a glass flask at 255-275°C. The degradation was studied by viscosity and COOH endgroup concentration. The concentration of carboxylic acid endgroups in PBT and $PBTA_{20}$ increased linearly with time. An autocatalyzing effect of the generated COOH endgroups was not observed. The degradation constants for PBT were comparable to the literature values. For $PBTA_{20}$ the degradation constant as measured with viscometry was the same as for PBT. The degradation constant of

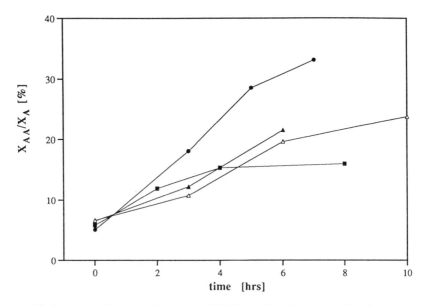

12. Formation longer amide units in PBTA$_{20}$ as function of reaction time :
■, 255°/vacuum; △, 270°/N$_2$; ▲, 270°/vacuum; ●, 275°/vacuum
(Reproduced with permission from reference **3**)

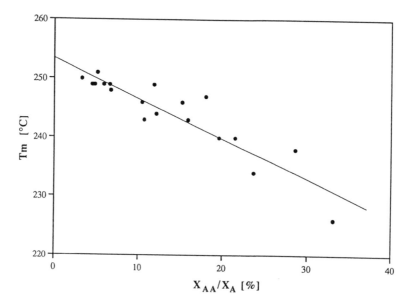

13. The melting temperature of PBT A$_{20}$ as function of the concentration of longer amide units

PBTA as determined from COOH concentrations was higher than for PBT and PBTA from viscometry.

With degradation time and temperature the diamide non uniformity increased. A loss of amide uniformity has a strong effect on the melting temperature of the PBTA. As a consequence, the purity of T4T determines its effectiveness in increasing the melting temperature of PBTA. If one wants to improve the properties of PBT with amides it is crucial that the amide segments are of uniform length.

Acknowledgments

This research was financially supported by GE Plastics (Bergen op Zoom, the Netherlands). J. Bussink and J. Feijen are acknowledged for the fruitful discussions and valuable suggestions. W. Lengton (University of Twente) is acknowledged for the development of a method for endgroup analysis.

References

1   Gaymans, R.J.; De Haas, J.L.;Van Nieuwenhuize, O. *J. Polym. Sci.: Part A: Polym. Chem.* **1993**, *31*, 575
2   van Bennekom, A.C.M.; Gaymans, R.J. *Polymer* **1996**, *37*, 5439
3   van Bennekom, A.C.M.; Willemsen, P.A.A.T.; Gaymans, R.J. *Polymer* **1996**, *37*, 5447
4   van Bennekom, A.C.M.; Gaymans, R.J. *Polymer* **1997**, *38*, 657
5   van Bennekom, A.C.M.; Pluimers, D.T.; Bussink, J.; Gaymans, R.J. *Polymer* **1997**, *38*, 3017
6   van Bennekom, A.C.M.; van den Berg, D.; Bussink, J.; Gaymans, R.J. *Polymer* **1997**, *38,* 5041
7   Ellis, T.S. *J. Polym. Sci., Polym Phys. Edn.* **1993**, *31*, 1109
8   Goodman, I.; Vachon, R.N. *Eur. Polym. J.* **1984**, *20*, 529
9   Goodman, I.; Vachon, R.N. *Eur. Polym. J.* **1984**, *20*, 539
10 Schmidt,B.; Koning, C.E.; Wrrumeus Buning, G.H.; Kricheldorf, H.R.;*J.Macromol. Sci.* **1997**,*A35,* 759
11 Bouma, K; Gaymans, R.J. *ACS Polym. Mat. Sci and Eng.* **1997**, *77*, 486,
12 Williams, J.J.R.; Laakso, T.M. *Brit. Pat.* **1995**, 824, 308
13 Stapert, H.R.; van der Zee, M.; Dijkstra, P.J.; Feijen, J. *ACS Polym.Mat. Sci and Eng.* **1997**, *76*, 414
14 Williams, J.J.R.; Laakso, T.M.; Contois, L.K. *J. Polym. Sci.* **1962**, *61*, 353
15 Della Fortuna, G.; Oberrauch, E.; Salvatori, T.; Sorta, E.; Bruzzone, M. *Polymer* **1977**, *18*, 269
16 Aharoni, S.M. *Macromolecules* **1988**, *21*, 1941
17 Serrano, P.J.M.; Thüss, E.; Gaymans, R.J. *Polymer* **1997** *38*, 3893
18 Serrano, P.J.M.; van de Werff, B.A.; and Gaymans, R.J. *Polymer* **1998**, *39*, 83
19 Serrano, P.J.M.; van Bennekom, A.C.M.; Gaymans, R.J. *Polymer*, in press
20 Pilati, F.; Manaresi, P.; Fortunato, B.; Munari, A.; Passsalacqua, V. *Polymer* **1981**, *22*, 1566
21 Passalacqua, V.; Pilati, F.; Zamboni, V.; Fortunado, B.; Manaresi, P. *Polymer* **1976**, *17*, 1044
22 Cognigni, F.; Mariano, A. (Anic.S.p.A., Italy), DE 3029970 A1, 1981
23 Pilati, F. *"Comprehensive Polymer Science"* , Ed G.Allen, J.C.Bevington, Pergamon Press, Oxford, 1989, Vol. 5, pp 201
24 Dinse, H.D.; Tuçek, E. *Acta Polymerica* **1980**, *31*, 108
25 Gaymans, R.J.; Amirtharaj, J.; Kamp, H.; *J. Appl. Polym. Sci.* **1982**, 27, 2513
26 Buxbaum, L.H., *J. Appl. Polym. Sci., Appl. Polym. Symp.* **1979**, *35*, 59

27 Schaaf, E.; Zimmerman, H.; Dietzel, W.; Lohmann, P. *Acta Polym.* **1981**, *32*, 250

28 Gaymans, R.J.; Schuyer, J. *"Polymerization reactors and Processes"* J.N.Henderson, T.C.Bouton Eds., Publisher ACS, Washington,Symposium Series , 1979, no.104, pp 137

29 US pat. **1996**, 5,496,887 (to BASF)

30 Gaymans, R.J.; Utteren, T.E.C.; van den Berg, J.W.A.; Schuijer, J. *J. Polym. Sci. , Polym. Chem. Edn* **1977**, *15*, 537

31 Deveaux, J.; Godard, P.; Mercier J-P. *Makromol. Chem.* **1978**, 179, 2201

32 Raju, K.V.S.N.; Yaseen, M. *J. Appl. Pol. Sci.* **1992**, *45*, 677

33 Rafler, G.; Blaesche, J.; Möller, B.; Stockmeyer, M. *Acta Pol.,* **1981**, *32*, 608

34 Pilati, F.; Masoni, S.; Fortunato, B. *Pol. Com.* **1984**, *25*, 190

Chapter 4

The Preparation and Melt Behavior of Oligomeric Bisphenol-A Polycarbonates

Joseph A. King Jr.[1]

Corporate Research and Development, General Electric Company, 1 Research Circle, Niskayuna, NY 12309

During the base catalyzed, neat condensation of diphenyl carbonate (DPC) with bisphenol A (BPA), a characteristic fluid-phase expansion of the molten BPAPC oligomers is observed. This liquid-phase expansion is indicated to be a state function of the reaction environment. The fluid behavior is strongly temperature and pressure dependent. Bubble formation/stability under the reaction conditions is related to the extent of reaction and the phenol velocity. The molecular weights of the BPAPC oligomers at this transition point, are found to be in the region were the melt behavior is changing from that of a viscous fluid to one of an entangled polymer

The excellent thermo-mechanical properties and clarity of bisphenol-A polycarbonate (BPAPC) have made it the subject of extensive research over the past few decades.[1,2,3] In particular, its high impact resistance over a broad temperature range below the glass transition temperature (T_g), great rigidity, and excellent dimensional stability allow for a variety of engineering applications. These mechanical properties, coupled with BPAPC's superb optical properties, have made it the material of choice for many glazing and magneto-optical applications.

A series of low molecular weight BPAPC oligomers were prepared, isolated and evaluated in an attempt to experimentally determine the entanglement molecular weight of the polymer. The entanglement molecular weight of BPAPC itself has been the subject of only a limited number of reports.[4,5] The presence of interchain entanglements are the putative requisite for a plateau modulus, G_N^o, in amorphous polymer systems.[6] In the viscoelastic region of neat, organic polymers, an inverse relationship between the plateau modulus and molecular weight between entanglements M_e (eq. 1) is commonly found.[5,6]

$$G_N^o \approx RT/M_e \qquad \text{eq. 1}$$

[1]Current address: GE Financial Assurance, 6604 West Broad Street, Richmond, VA 23230.

For pure polymers in their molten state, the log-log plot of the M_W versus their zero-shear viscosity (η_o) generates an inflection point in the resultant curve. This inflection point is taken to indicate where the molecular weight dependence changes from a simple, linear $M_W^{1.0}$ to a $M_W^{3.4}$ dependence.[6] The molecular weight of the polymer at this inflection point in the log-log plot is referred to as the characteristic molecular weight for entanglement, M_C. The reported (calculated) value for BPAPC is $M_C = 4875$.[4,5] This critical molecular weight for entanglement was derived from an evaluation of the plateau region in its dynamic mechanical spectrum for high molecular weight BPAPC ($M_W > 20,000$). For measured polymeric systems, $M_C \approx 2M_e$.[7,8] Thus, it is possible to derive the M_C due to an existing relationship with the plateau modulus which can be approximated using equation 2:

$$G_N^{\,o} \approx 2RT/M_C \qquad \text{eq. 2}$$

Efforts to confirm the postulated value through the direct synthesis of low molecular weight oligomeric materials ($M_W = 2000\text{-}8000$) have not been reported. The paucity of studies for oligomeric BPAPC may be the result of the limited number of simple, selective synthetic methods for their preparation.[9]

We report here both the facile synthetic preparation of oligomeric BPAPC as well as their general melt viscosity behavior.

Experimental Methods

The oligomeric polycarbonates studied were prepared via reported melt polymerization procedures.[10] A number of the materials were precipitated from CH_2Cl_2/THF to facilitate the light scattering and viscotex measurements. The following procedure for precipitation was used: 2.00g of material dissolved into 20ml of methylene chloride. The resulting solution was poured into 100ml of acetonitrile. The white precipitate was collected by gravity filtration and then dried for 24h at 15mmHg and 70°C; when larger quantities of materials were required this procedure could be scaled accordingly. All samples were run in triplicate with a majority of the measurements being the average of five runs.

The light scattering system was equipped with an HP model 1050 pump, HP model 1050 autosampler, HP model 1047A refractive index detector, and a LDC model KMX-6 light scattering detector. The apparatus used a bank of three micro styragel columns having pore sizes of 10^5Å, 10^3Å, and 500Å. The solvent was 1% ethanol in chloroform at a flow rate of 1.5ml/min. The column temperature was 25°C. The LALLS data was collected and processed using Viscotex GPC-LS software.

The preliminary (screening) zero-shear viscometry measurement were performed using a Brookfield Thermosel Digital Viscometer Model # LVTDV-II under low (zero shear) conditions; the shear rates were 0.396-15.84 sec^{-1}. Due to the fixed volume of the heated thermo-container cavity, each sample mass evaluated was dependent on the spindle used: 10g for spindle 18 and 18g for spindle 34. Spindle 34 was used for the higher viscosity oligomer melts. The Brookfield response behavior was verified using both Brookfield and SF96 standards at 25°C. These standard viscosities spanned a range of 50-13,000 cps. The measured deviation in viscosity for spindle 18 from the reported values averaged 1.5% lower than the standards. Spindle 34 averaged 7-9% higher values than spindle 18.

Prior to both the preliminary analytical analysis and rheology disk pressing, all samples were dried (80°C) under vacuum (~15mmHg) over night. After pressing the samples into 25mm diameter disks the samples were redried overnight as before; using pre-heated Teflon faced 304 stainless steel frame molds, the samples were partially melted in the press for 2-3 min. (445°F; ~228°C) without pressure, then 2.2 psi of pressure was added for 2-3 min., and finally raised to 13.23 psi for 3-4min..

The time sweeps were performed on a Rheometrics Dynamics Spectrometer (RDS) 7700. The spectrometer was configured with 25mm diameter parallel plates and a nitrogen purged oven. This yielded sample sizes of roughly 25mm diameter x 2mm thick; the material mass evaluated was slightly greater than one gram. An experimental temperature of 300°C was used in all the rheological tests. The melt stability measurements were performed at a frequency of 1 rad/s.

The GPC samples were run on a HP 1090L Liquid Chromatograph modified for GPC work. The GPC used an HP autosampler and 254nm UV detector. The GPC used a bank of two ultra styragel linear columns having a pore size range of 10^5Å to 500Å. The GPC was calibrated using narrow molecular weight polystyrene standards and the results reported are relative to polystyrene. The solvent was pure chloroform at a flow rate of 1.0ml/min. The columns were held in heated, constant temperature oven at 40°C. The GPC data was collected and processed using a PE Nelson model 2600 data system.

The T_g's were determined on a Perkin-Elmer DSC-7 configured with a two-stage intracooler and was purged constantly with dried, research-grade nitrogen. The instrument was calibrated at a heating rate of 20°C/min. using a 99.999% pure indium standard. The oligomeric powders were weighed to the nearest tenth of a milligram and crimped into aluminum sample pans (TA Instruments). The samples were each placed in the instrument at 25°C.

The IV's (dL/g) were run in chloroform at 25°C using an Ubbelohde Viscometer apparatus and standard dilution methods: $(T - T_0)/(T_0 \times C) = R_{vis}$ where T = sample flow time, T_0 = pure solvent flow time, C = sample concentration, and R_{vis} = reduced viscosity. Each run was determined at three concentrations. The reduced viscosity is plotted versus concentration to obtain the intrinsic viscosity (y-intercept). Each reported value is the average of three runs.

Statistics[10,11] - Set-up of the experimental designs and analysis of the data was done using the RS/1 - RS/Explore - RS/Discover software series from BBN Software Products, Cambridge, MA. The experimental data was analyzed using the multi-factor regression techniques in the RS/Explore MULREG environment.

The data was transformed to an orthogonal scale by subtracting the mid-point and dividing by the range of each factor. This is standard procedure when it is suspected that the intersection and quadratic terms may be present. The multiple regressions were examined using ANOVA tables and graphical tools. Residuals were examined for outliers using normal probability plots. Once the data were cleaned, the model was refined using stepwise regressions, removing all non-significant terms. There was a clean break between significant and non-significant terms in most of the models, with the significant terms having "P-to-center" <0.01 and the non-significant terms >0.20.

Visualization of the resulting models was done with contour plots and prediction plots. The prediction plots include confidence intervals; 95% simultaneous intervals were used in this program. The confidence intervals aid in estimation of real differences.

The laboratory melt polymerizer was outfitted with a Shimpo Ringcone metallic traction drive motor unit model NRX attached to a Cosmic Industries Co., Ltd. motor cylinder model BJ-100K 3JPL.

A standard oligomerization reaction was run using the following general program to facilitate the selective distillative removal of phenol: i) 210°C at 200mmHg for 35min; ii) 210°C at 100mmHg for 35min; iii) 240°C at 15mmHg for 40min and: iv) 270°C at 2mmHg for 20min. The reactor vessel was loaded with diphenyl carbonate (DPC; 138.9g; 0.648mol), BPA (136.8g; 0.600mol), and tetramethyl ammonium hydroxide (0.027g; 0.0003mol). The solution was stirred at 250rpm throughout the course of the reaction. The reaction times and/or temperatures were varied occasionally to evaluate their influence on the on-set or termination of the molten fluid expansion. The torque remains constant during the reaction until roughly 105-110mls (112-118g) of material has distilled from the reactor vessel; the theoretical amount of pure phenol distillate is 112.9g based on the moles of BPA. A slight excess of DPC (3-20mol%) is generally used which makes up the bulk of the excess greater-than-theoretical distillate. The various distributions of oligomers were obtained by terminating the reactions at different points determined by either of two (related) methods: 1) stopping at a pre-determined distillate volume within the range 104-119ml or 2) at various degrees of expansion of the fluid phase. The melt samples were then precipitated using the above procedure.

* The following examples are representative of the general melt foaming experimental procedure which are also listed in **Table II**. *

DPC-BPA Ratio = 1.000 for Formation of BPAPC Resin (Procedure A):
BPA (136.9g; 0.600mol) and DPC (128.6g; 0.600mol) were added into a one liter glass melt polymerization reactor as powders; the glass reactor surfaces had been previously passivated via acid washing, rinsing, and subsequently drying at 70°C over night. A solid nickel, right-handed, helixing stirrer was suspended in the powder and the reactor vessel/stirrer unit attached to the melt apparatus. The reactor vessel was deoxygenated by evacuation to about 1 torr and then refilling the vessel with purified nitrogen. This deoxygenation procedure was repeated a total of three times. The reactor vessel was immersed in a fluidized heat bath preheated to 180°C. The reaction mixture was allowed to melt, producing a colorless, homogeneous liquid; once a majority of the solid material melts, the remaining powder suspension can be slowly stirred to promote better heat exchange. Upon complete solution, the

system was allowed to thermally equilibrate (5-10min). The solution was stirred at 250rpm. Into this solution was syringed the base catalyst: $TMAH_{aq}$ (3 x 10^{-4}mol; 5.0 x 10^{-4}mol/molBPA equivalents) and $NaOH_{aq}$ (300μL of a 0.025M solution; 7.5 x 10^{-6}mol); $TMAH_{aq}$ is an aqueous solution of tetramethylammonium hydroxide. The resulting solution was stirred at $180^{o}C$ for 5min. At this time the reaction temperature was raised to $210^{o}C$ and the pressure lowered to 175mmHg. Phenol began to distill from the reactor immediately (approx. 3-4 drops/sec.). After 35min, the reactor pressure was lowered to 100mmHg and held there for an additional 35min. Phenol continued to distill into the receiver flask (1-2drops/sec) during this time and a total volume of 74mL was collected by the end of $210^{o}C$ stage. The reactor temperature was raised to $240^{o}C$ (15torr) and these conditions maintained for 20min. During this time period, phenol distilled at an average rate of about 1drop/3-5sec (a total of 95mL were collected to this point). The reaction temperature was held at $240^{o}C$ (5torr) for 60min. The reaction was terminated. A total of 108.5 grams of distillate was collected during the course of the reaction. The colorless oligomeric polycarbonate was collected yielding the following analytical data: $IV_{chloroform}$ = 0.126, M_w = 5,101, M_n = 2,623, M_w/M_n = 1.945, M_z = 8,297, and OH = 0.974wt%.

DPC-BPA Ratio = 1.005 for Formation of BPAPC Resin (Procedure C):

BPA (136.8g; 0.600mol) and DPC (129.2g; 0.603mol) were added into a one liter glass melt polymerization reactor as powders; the glass reactor surfaces had been previously passivated via acid washing, rinsing, and subsequently drying at $70^{o}C$ over night. A solid nickel, right-handed, helixing stirrer was suspended in the powder and the reactor vessel/stirrer unit attached to the melt apparatus. The reactor vessel was deoxygenated by evacuation to about 1 torr and then refilling the vessel with purified nitrogen. This deoxygenation procedure was repeated a total of three times. The reactor vessel was immersed in a fluidized heat bath preheated to $180^{o}C$. The reaction mixture was allowed to melt, producing a colorless, homogeneous liquid; once a majority of the solid material melts, the remaining powder suspension can be slowly stirred to promote better heat exchange. Upon complete solution, the system was allowed to thermally equilibrate (5-10min). The solution was stirred at 250rpm. Into this solution was syringed the base catalyst: $TMAH_{aq}$ (3 x 10^{-4}mol; 5.0 x 10^{-4}mol/molBPA equivalents) and $NaOH_{aq}$ (300μL of a 0.025M solution; 7.5 x 10^{-6}mol). The resulting solution was stirred at $180^{o}C$ for 5min. At this time the reaction temperature was raised to $210^{o}C$ and the pressure lowered to 200mmHg. Phenol began to distill from the reactor immediately (approx. 3-4 drops/sec.). After 35min, the reactor pressure was lowered to 150mmHg and held there for an additional 35min. Phenol continued to distill into the receiver flask (1-2drops/sec) during this time and a total volume of 65mL was collected by the end of $210^{o}C$ stage. The reactor temperature was raised to $240^{o}C$ (100mmHg) and these conditions maintained for 10min. The pressure was then lowered to 50mmHg and held there for 20min. The pressure was further reduced to 15torr and held for 20min. Up to this point, a total of 102mL were collected. The reaction temperature was held at $240^{o}C$ (5torr) for an additional 60min and then the reaction was terminated. A total of 107.6 grams of distillate was collected during the course of the reaction. The colorless oligomeric polycarbonate was collected yielding the following analytical data: $IV_{chloroform}$ = 0.121, M_w = 4,758, M_n = 3,128, M_w/M_n = 1.521, M_z = 6,757, and OH = 1.010wt%.

Cautionary Note: Hot phenol is evolved (distilled) during the condensation polymerization of BPA with DPC. Hot phenol readily attacks exposed flesh and is listed as a toxic substance. Special handling precautions should be used such as the wearing of chemically resistant gloves, face shields, *etc.* when working around this material.

Results and Discussion

The various BPAPC oligomers were prepared and isolated using a neat, melt polymerization procedure. This process involved the condensation of diphenyl carbonate with bisphenol A in the presence of a basic catalyst.[10] The point of oligomer collection was selected based on the occurrence of a melt event. Towards the latter stage of the reaction, the oligomeric liquid is observed to undergo a transient fluid expansion. The oligomers were isolated around where the molten fluid is undergoing the liquid phase transition. This transition manifests itself by a net expansion of the molten fluid. The fluid expansion results from stable bubble formation which propagates through the entire solution. This transitional expansion or "foaming" process persists for roughly 5-18 minutes depending on the reaction conditions.

The occurrence of this event is critical to the development of high molecular polymer. When this event does not occur, neither high quality nor high molecular weight polymer is observed; the torque on the stirrer necessary to maintain constant rpm's increases rapidly once the transition is nearly complete.[13] The foaming event is shown sequentially in **pictures 1-8**. The first picture shows an initial change in solution opacity due to micro-bubble formation. **Pictures 2-4** show the presence of the "foam" on the surface of the molten resin followed by its downward propagation through the solution until the entire melt is foamed. **Pictures 5-8** indicate that, once maximum expansion is reached, the foam slowly collapses back to a viscous, transparent fluid.

The selection of an on-set point for foam initiation is arbitrary. Our definition requires a net fluid expansion of the melt. Prior to the observance of the foaming transition, the melt undergoes a noticeable change-in-opacity (**picture 1**). This hazy or cloudy appearance is due to the formation of micro-bubbles in the solution (**figure 1**). Once bubble formation becomes stable - due to an increase in the melt strength of the medium - it propagates slowly downward through the melt. Hence, the difference between a "foaming" and a "non-foaming" melt event is not merely the presence of bubble formation but rather bubble formation with a net increase in the total volume of the liquid phase. The downward propagation of the foam through the solution would be anticipated to occur in most laboratory or CSTR type reactors: the growth of the BPAPC oligomers is fastest at the melt surface due to a more rapid loss of

Picture 1

Picture 2

Picture 3

Picture 4

Picture 5

Picture 6

Picture 7

Picture 8

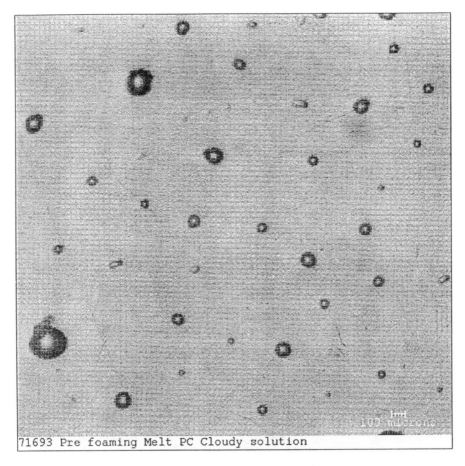

71693 Pre foaming Melt PC Cloudy solution

Figure 1:   Pre-Foaming Solution Micro-Bubbling

phenol at the melt interface. This mechano-molecular bubble formation-propagation event occurs in each and every melt reaction which is run to high molecular weight.

For sample analysis, aliquots were taken at the on-set, maximum extent of fluid expansion, and cessation of the melt transition. These materials were then used to correlate the dynamic melt behavior with the BPAPC molecular weight and the extent of reaction - from the phenol evolved. Little work on the characterization of ultra-low molecular weight BPAPC has been reported. Although nothing unusual was anticipated, some effort was spent correlating GPC and viscotex data with light scattering data to validate this assumption; to facilitate the absolute molecular weight light scattering measurements, a number of the samples were reprecipitated to produce narrower oligomer dispersivities (1.1-1.3). The molecular weight correlations indicate a simple linear relationship exists between both the light scattering vs Viscotex molecular weights (slope = 1.03; r = 0.997; 10 points) and the GPC vs Viscotex molecular weights (slope = 0.511; r = 0.990; 18 points).

A molecular weight comparison of the oligomers produced in the foaming régime was evaluated (**Table I**). The data indicate a GPC weight response relative to their absolute M_w similar to that observed for normal BPAPC molecular weight grades (0.511). The GPC molecular weight data (polystyrene standards) for the oligomers are a factor of 1.957 higher than via a viscotex/absolute method in chloroform solvent. A number of the viscotex values were themselves checked using laser light scattering to verify the consistency of their absolute molecular weights. The factor of two difference between the $GPC_{polystyrene}$ based and absolute molecular weights from light scattering measurements has been known for some time.

Table I: Representative Light scattering versus Viscotex and GPC BPAPC Oligomer Molecular Weights

Reaction Sample	Temp/Press.	Light Scattering Mw	Viscotex Mw	GPC Mw
Pre-foaming	240°C/100torr	2248	2230	3920
Pre-foaming	240°C/15torr	2516	2471	4730
On-set	240°C/15torr	3953	3874	7160
On-set	270°C/2torr		4365	7810
On-set	240°C/50torr		4646	9240
Pre-foaming	240°C/100torr	4472	4685	9070
Early Foaming	240°C/15torr		4856	8690
On-set	300°C/2torr	5206	5391	10030
Foaming On-set	270°C/15torr		5593	10910
Foaming On-set	300°C/50torr		5903	11480
Early Foaming	270°C/15torr	5981	5927	10910
Maximum Foam	240°C/15torr		5952	11540
Foaming On -set	300°C/15torr		6828	11920
Pre-foaming	300°C/100torr		6913	13550
Maximum Foam	240°C/5torr	7129	7453	14280

The correlation of the glass transition as a function of molecular weight was examined to ensure nothing unusual occurs in low molecular weight BPAPC. The plot of the oligomeric resin's Tg (^{o}C) versus molecular weight (Mw) is shown in **figure 2**. Nothing extraordinary is observed, which is what one would expect for the oligomers.

End-group Effects

The influence of the DPC/BPA ratio (potential end-group effects) could be evaluated directly in the laboratory. Since the transition process was believed to be a state function, no end-group dependence was anticipated. As expected, there is no end-group effect on the on-set of foaming. This is best illustrated by the time-to-foam data in **Table II**. These data indicate that, for a given set of reaction conditions, the times for the on-set of foaming are roughly identical. This behavior results because the time-to-foam data - for any given set of processing conditions - tracks linearly with the extent of reaction (*i.e.* loss of phenol), not the total mass loss or type of end-group (OH vs. phenyl). The laboratory batch reactor system is run under chemical equilibrium where distillative mass loss is rate limiting at the lower melt viscosities.

Molecular Weight (Mw) vs Polycarbonate Tg

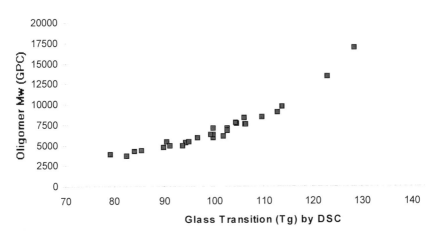

Figure 2: The glass transition of oligomeric BPAPC versus their Molecular weight

Only in cases of very high DPC loading do anomalously disparate on-set times occur. This anomalous behavior results when the large excess of free, residual DPC lowers the reaction melt viscosity by both disrupting the degree of polymerization (equilibrium) yielding a population of lower molecular weight oligomers per unit time. Thus, as defined by the reaction conditions, the on-set of reactor foaming is indicated to be a simple function of the extent of conversion.

Table II: Rate comparison for oligomeric and polymeric BPAPC build.

DPC/BPAs	Condition	Time*	IV (dL/g)	Mw	Mn	Mw/Mn	Mz	OH (wt%)
1.000	A	150min	0.126	5,100	2,620	1.945	8,300	0.974
1.150	A	150min	0.131	6,300	3,190	1.977	10,280	0.481
1.250	A	150min	0.134	5,730	2,990	1.920	9,270	0.307

DPC/BPAs	Condition	Time*	IV (dL/g)	Mw	Mn	Mw/Mn	Mz	OH (wt%)
1.005	C	180min	0.121	4,760	3,130	1.521	6,760	1.010
1.125	C	180min	0.124	5,210	2,740	1.905	8,390	0.547
1.250	C	180min	0.119	4,900	3,250	1.508	6,930	0.261

conditions:

A = 210C/175mm(35min)-210C/100mm(35min)-240C/15torr(20min)-240C/5torr(until foaming)

C = 210C/200mm(35min)-210C/150mm(35min)-240C/100mm(10min)-240C/50mm(20min)-240C/15torr(20min)-240C/5torr(until foaming).   * Time = Total time till the on-set of foaming.

Pressure Effects on the Fluid Expansion

The melt foaming process is characterized by the net expansion of the polymer melt. The maximum height or degree of expansion of the foam is nearly linearly related to the decrease in reactor pressure (**Graph 1**). The data are plotted for the resin volume (100 = initial or normal liquid mass-volume at temperature) versus the reactor temperature and pressure. **Graph 1** indicates that, once a threshold pressure has been breached, the resin expansion increases as an inverse function of the pressure. Below the pressure threshold limit, there is an observable expansion of the melt fluid during the viscous-liquid to polymer melt transition. Above this threshold pressure limit there is no detectable net fluid expansion, yet micro-bubbling and "normal" surface boiling may be evident.

Consistent with the idea that the "observance" of melt foaming is a state function, a number of experiments were performed where the pressure was varied dramatically once foaming had begun; usually the pressure was raised from an initial value (<<75mmHg) to ≥100mmHg, held at the higher pressure for variable periods of time, then returned to its foaming value. During these excursions, the melt went from a well behaved, stable foam to a relatively placid, viscous liquid as the pressure was raised. Upon returning the system below the threshold value, foaming resumed. At a given temperature, this pressure dependent foaming/non-foaming process can be cycled repeatedly provided the extent of reaction is not allowed to progress substantially. **Table I** includes examples (240°C/100torr and 300°C/100torr) where the pressure was kept at 100torr throughout the latter stage of the reaction. Their molecular weights are in the range were foaming should occur but isn't observed; this is consistent with the need for a high vapor velocity for foam formation. These systems did continue to build molecular weight but extremely slowly. The listed

Graph 1: Temperature and Pressure Effect on Foam Expansion
MAXFOAM

FTEMP

———···——— MFOAM

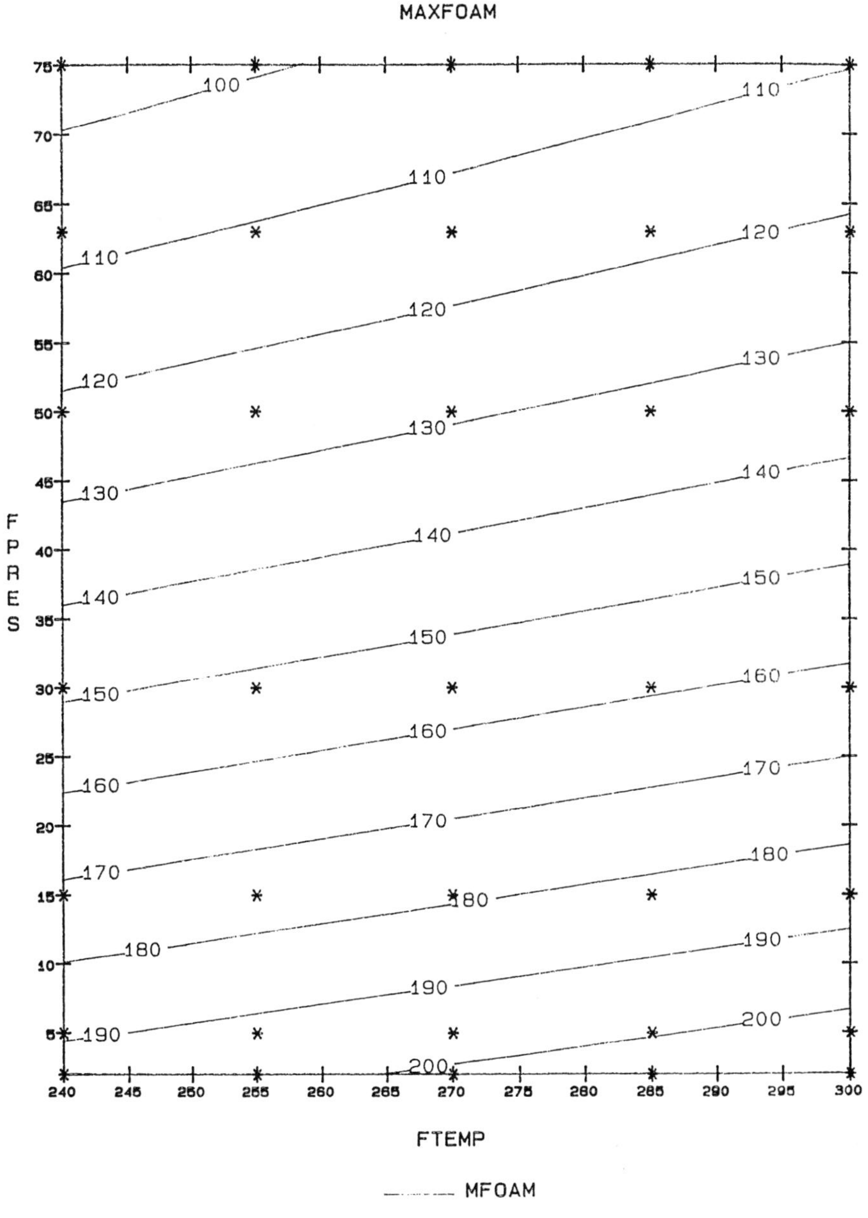

values are at four (first 240°C/100torr) and six hours (second 240°C/100torr and 300°C/100torr). The other listings in the table measured at between 30min and 1hour.

Temperature Effects on the Fluid Expansion

The effect of temperature on fluid expansion is much less significant. The incremental expansion of the melt fluid due to an increase in the reactor temperature is only an additional 5-8% of the foam's overall expansion in the 240-300°C range. For comparative purposes, the density of the BPAPC oligomeric melts change only ~8.5% (due to liquid expansion) over the same temperature range at atmospheric pressure and constant IV (**Graph 2**). Thus, the observed temperature dependent component of the foam expansion may be attributable to merely the temperature dependent change in fluid density and associated melt viscosity.[1] The strong pressure and mild temperature dependence is what one would expect. For the temperature ranges examined, once significantly above the boiling point of the blowing agent (bp$_{PhOH}$ = 180-182°C), the vapor load and velocity are a stronger function of pressure than temperature.

The melt viscosity of the on-set reactions was checked over the range from 240-300°C. The melt viscosity versus molecular weight (GPC) at 300°C is shown in **figure 3a**. The log-log plot of these data at 240, 270, and 300°C plot as straight lines with a slope of ~1.0 (**figure 3b**: at 300°C). The slope's value is consistent with the dominant material flow property being similar to that of a viscous fluid. This is expected since the critical molecular weight for entanglement (M_c) is at the upper end of these oligomer molecular weights (**figure 4**); the on-set M_w vs. IV data have been

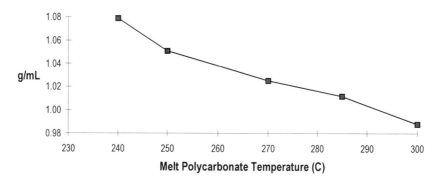

Graph 2: BPAPC Density Temperature Dependency

g/mL

Melt Polycarbonate Temperature (C)

corrected to read in absolute M_w. At lower temperature, the % entanglement necessary to produce the required melt strength to sustain bubble formation would be less than at higher temperatures.[14]

The melt viscosity at the on-set of foaming versus intrinsic viscosity is shown in **figure 5**; the melt viscosity data here are reported as measured at the foaming reactor temperature. The data span a fairly narrow range of IV's and melt viscosities.

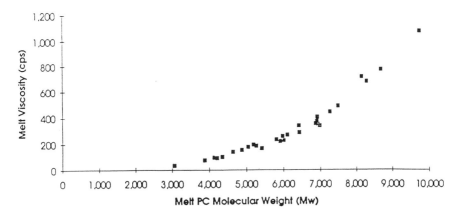

Figure 3a: Melt Viscosity versus Molecular Weight at 300°C

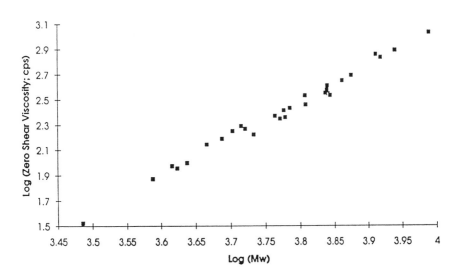

Figure 3b: Melt Viscosity versus Log(Molecular Weight) at 300°C

66

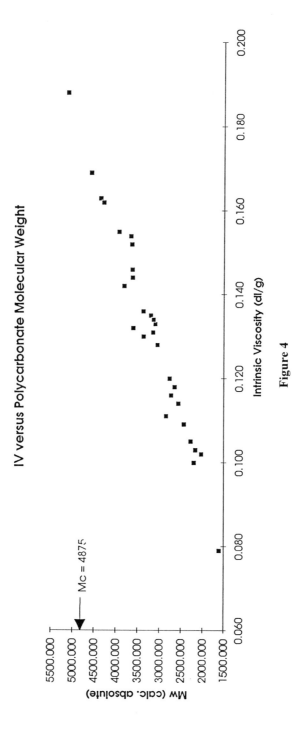

IV versus Polycarbonate Molecular Weight

Mc = 4875

Mw (calc. absolute)

Intrinsic Viscosity (dl/g)

Figure 4

Similarly, when one compares the (at temperature) on-set melt viscosities of the oligomers versus their molecular weight, a narrow dispersion is observed (**figure 6**).

DOE Evaluation[15]

The experimental design for sample collection and expansion measurement were structured for statistical evaluation using standard methods of Design of Experiments (DOE). The data was then analyzed using standard regression analysis. The reactor process data are divided into two parts. The first part relates to the conditions and compositions present at the (initial) on-set of the foaming phenomenon. The second part of the data are taken from the material at the cessation of foaming; *i.e.* once the foam collapses back to a transparent, colorless, viscous melt. These two sets of data can be used to bracket the reactor foaming régime. This bracketing is important for predicting the occurrence of the transition since the foaming phenomenon is a state function which is independent of the reactor type or design.

Stirring Effects

Two types of reactor stirrers were used to evaluate the effect of mixing efficiency on foaming.[16] The relative statistical influence of pressure, temperature, and stirrer design is shown in the **Pareto Graph**. No stirrer dependent effects (F-ratio = 0.08)[17] were indicated in the statistical analysis.

Foaming On-set Intrinsic Viscosities and Molecular Weights Data

The on-set data indicate a strong positive (higher) dependence of the M_w and IV on temperature and a weaker negative (lower) dependence on pressure; these effects are only detectable once the pressure is below the threshold limit. The F-ratios of 141 and 20, respectively, indicate that the significance of the temperature and pressure effects are >99.99%. The refined models for both IV and M_w have only the linear temperature and pressure terms, and their R-squared values are 0.799 and 0.780, respectively. These indicate that both models are "explaining" about 80% of the variation in the data, with the rest of the variation attributable to random effects.[18] In the analysis of the on-set data, one outlier (out of 48 total points) was removed from the analysis; thereafter the residuals showed no other unusual properties.

The foaming on-set IV's versus temperature and pressure are plotted in **Contour plot 1a**. The corresponding molecular weight dependencies (GPC$_{Polystyrene}$) are shown in **Contour plot 1b**. The straight lines of these plots are indicative of the linearity of the effects. The predictions of the IV contour plot **1a** are ±0.005 IV units across the entire range.[15] **Contour plot 1b** has a confidence range around its predictions which varies from ±210 M_w units at its center to ±400 M_w units at its edge.

Foam Cessation IV's and M_w Data

The foam cessation data indicate a strong temperature dependence. However, the data do indicate a substantial and much stronger pressure dependence than in the on-set data. There are non-linearities in both the IV and M_w data, as well. The ANOVA analysis shows that the pressure and temperature effects are about equally significant (F-ratios of 256 and 303, respectively). The non-linear effects are real but

68

Foaming: Intrinsic Viscosity versus Melt Viscosity

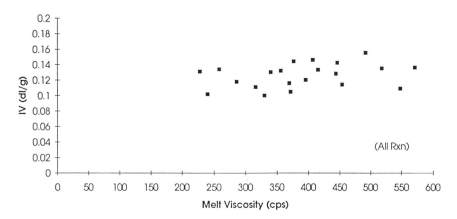

Figure 5

Laboratory Melt Reactor Foaming
(All Temperatures and Pressures)

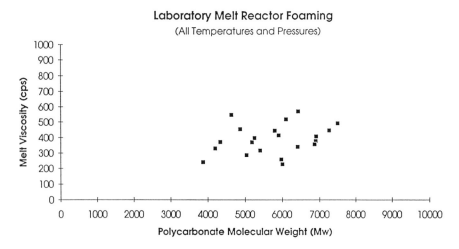

Figure 6

Pareto Graph

Pareto Graph for Mulreg OLD_NEW_STIR_MUL, Model OLD_NEW_STI
Main Effects on Response MW

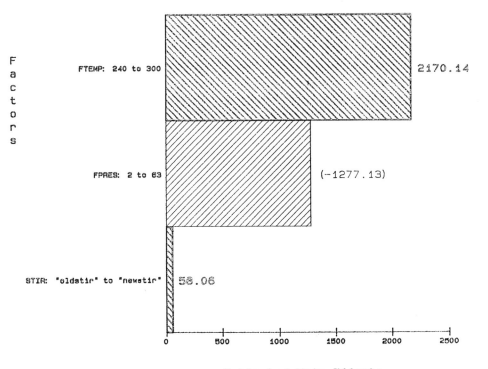

Estimated Main Effects

70

Contour Plot 1a: Foam On-set IV's versus Temperature and Pressure

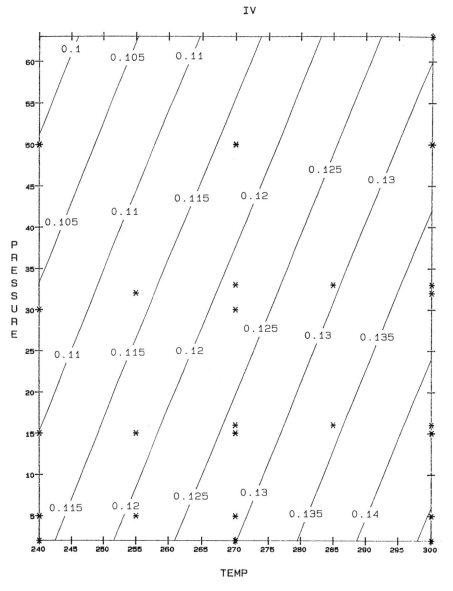

IV

TEMP

————— IV

Contour Plot 1b: Foam On-set M_w's versus Temperature and Pressure

MW

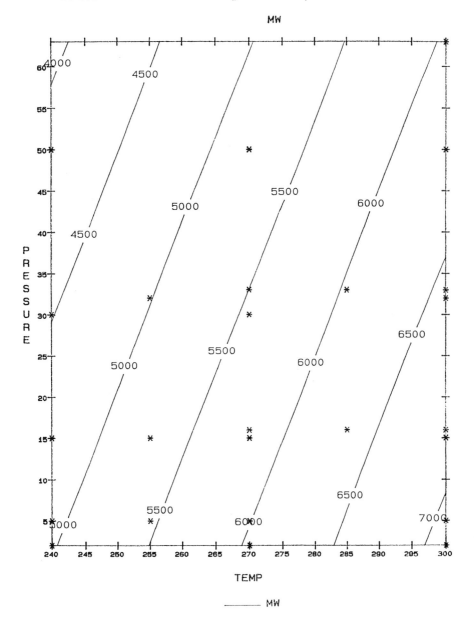

TEMP

——— MW

small. The R-squared values for both the IV and M_w models are 0.95, indicating that they have a very high degree of fit to the data. No outliers were removed and no unusual properties were observed in the residuals.

The contour plots for both the foam collapse IV's versus temperature and pressure and for the molecular weight dependency are shown in **Contour plots 2a & 2b**, respectively. Both plots indicate that in the lower pressure range (reactor pressure < 30mmHg) the effects of temperature and pressure are relatively linear; it is at the higher pressure range that they exhibit a strong non-linearity. For this reason, the confidence intervals increase from ±0.02 IV units at lower pressure to ±0.04 IV units at higher pressure and temperature. Similarly, the molecular weight confidence interval (GPC) increase from ±1000 M_w units at lower pressure to ±2600 M_w units at higher pressure and lower temperature.

Molecular Rationale

There appears to be a molecular basis for the transient occurrence of the melt condensation polymerization foaming. The absolute molecular weight range for the BPAPC oligomers collected in the foaming regime transcends the calculated critical entanglement molecular weight for BPAPC.[4,5] Again, for pure polymers in their molten state, the log-log plot of the M_W versus the zero shear viscosity (η_o) should generate an inflection point in the resultant curve. The inflection point occurs in the region where the molecular weight flow behavior changes from that of a simple, linear dependence ($M_W{}^{1.0}$) to a 3.4 power dependence ($M_W{}^{3.4}$).[6] The reported (calculated absolute) value for BPAPC is $M_C = 4875$.[4,5] It is this entanglement phenomenon which imparts sufficient melt strength to the molten resin to sustain the observed bubble formation and, hence, stable foaming propagation.

The extent of reaction for the on-set of stable bubble formation is strongly temperature and pressure dependent; at lower temperatures (220-240°C), a lower entanglement percentage is required to attain adequate melt strength, than at higher temperatures. The extent of reaction (ρ) at the transition on-set for the 240°C reactions is roughly $\rho = 0.908$-0.924, while at 300°C, it is approximately $\rho = 0.933$-0.947. From these data, the approximate degree of polymerization is $DP_{240C} = 10.9$-13.2 and $DP_{300C} = 14.8$-19. For comparison, the degree of polymerization where $M_c = 4875$ would be DP = (4875/254) = 19.2. These data indicate the melt foaming phenomenon occurs in a region were appreciable amounts of linear, entangled BPAPC polymer is predicted to exist. The observed bubble stability and propagation are consistent with a change in behavior of the molten oligomers from that of a simple, viscous fluid to one of an entangled polymer

Experimental observations are consistent with a degree of entanglement transition and its associated large change in solution melt viscosity. There is never an apparent increase in torsional strain on the motor-stirrer assembly of the laboratory melt polymerizer prior to the occurrence of foaming.[19] Roughly halfway through the foaming stage the torsional strain on the motor begins to increase quite dramatically to between 20-200 poise/min.

Since a Boltzmann distribution of molecular species results from an AA + BB condensation polymerization, only small amounts of entangled species may be necessary to obtain sufficient melt strength to sustain a foaming state. This is similar

Contour Plot 2a: Foam Collapse IV's versus Temperature and Pressure

IV

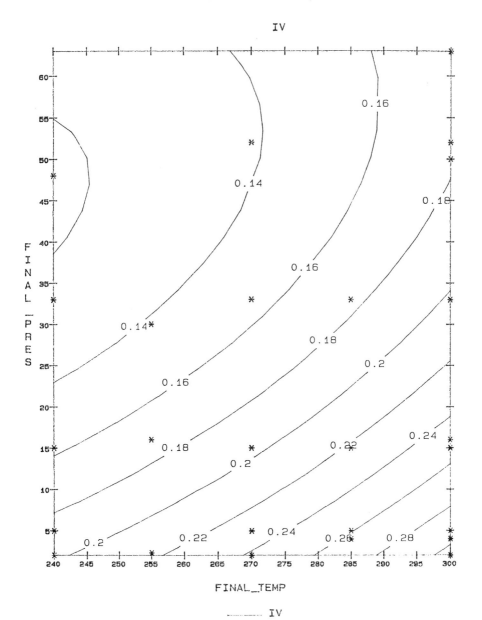

FINAL_TEMP

———— IV

Contour Plot 2b: Foam Collapse M_w's versus Temperature and Pressure

MW

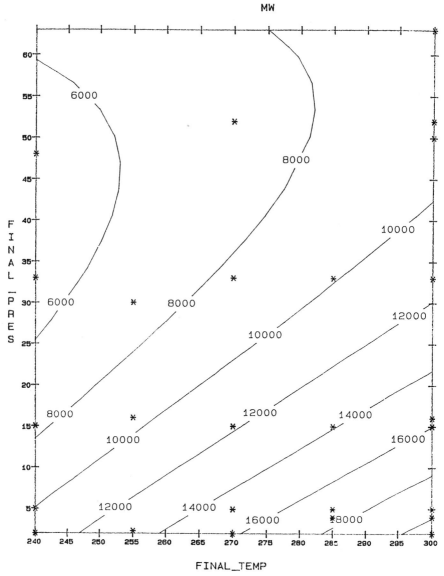

FINAL_TEMP

——— MW

to the effect observed in branched blow molding resin grades where small amounts of a branching agent (0.2-0.8wt% - increases entanglement density) impart a substantial amount of melt strength. If we assume the system is at chemical equilibrium with a most probable distribution of molecular weights, the percent of entanglements present may be calculated. Thus, the number necessary to obtain the requisite melt strength for the on-set of bubble stability may be estimated.

This melt transition behavior is not limited merely to the BPA homo-polymer. The transitional expansion or foaming process is common to most other condensation polymerization reactions; the exceptions to this rule would be either systems which don't generate blowing agents - such as cyclic material condensations - or those of limited or no entanglements such as in liquid crystalline materials (no real M_c). Any homo- or co-polymer reaction which liberates a blowing agent under conditions where its vapor velocity is high enough to generate bubble formation, will exhibit this phenomenon. The foaming transition will occur where the melt strength is sufficient to sustain bubble formation. This viscosity range is between 200-800cps with 300-600cps being the more usual. Under normal melt processing conditions, the on-set of this melt viscosity range generally occurs where the population of entangled polymer ranges from 0.5-15wt%; the "normal" melt conditions will be different for each class of reactions as well as compositionally or class specific. Thus, melt foaming is commonly observed in polycarbonates, polyestercarbonates, polyesters, polyesteramides, polyurethanes, polyureas, and polyamide-carbonates. Certain types of polyimide, polyetherimide, polyamides and silicone condensation systems should exhibit this transition, as well.

Summary

During the base catalyzed, solventless condensation of diphenyl carbonate with bisphenol A, a characteristic fluid-phase transition of the molten BPAPC oligomers is observed. This liquid-phase expansion is indicated to be a state function of the reaction environment. The degree of fluid expansion is a strong function of pressure. The principle blowing agent for the expansion-bubble formation is phenol; cessation of bubble formation results from the diminished evolution of phenol due to the high extent of conversion of the monomers. The general on-set and cessation of the BPAPC melt reactor foaming phenomenon is bracketed using the process variables of temperature and pressure. The molecular weights of the BPAPC oligomers at the transition point, are found to lie in the region were the melt behavior is changing from that of a viscous oligomeric fluid to one of an entangled polymer. The experimentally determined entanglement molecular weight from this work is consistent with the numerically calculated value derived previously from the plateau modulus.[4,5]

Acknowledgments

The author would like to thank Jim Cawse for his superb DOE assistance, Jim Kelly for the rheological work, Sue Weissman (IV), Linda McCracken (T_g; DSC), and David Dardaris (GPC) for their supporting analytical work. The author would also like to thank Godavarthi (Raj) Varadarajan and Bill Richards for helpful discussions concerning this phenomenon.

References

1. Schnell, H., "The Chemistry and Physics of Polycarbonates"; Interscience Publishers, John Wiley & Sons, Inc., New York, 1964, 1.
2. Fox, D. W., "Polycarbonates"; in Kirk-Othmer Encyclopedia of Chemical Technology, 3rd Edition, Wiley, New York, 1982, Vol. 18, 479.
3. Fritag, D., Fengler, G. and Morbitzer, L. *Angew. Chem. Int. Ed. Engl.* **1991**, *30*, 1598.
4. Aharoni, S.M. *Macromolecules* **1983**, *16*, 1722.
5. Aharoni, S.M. *Macromolecules* **1986**, *19*, 426.
6. Ferry, J.D. "Viscoelastic Properties of Polymers"; John Wiley & Sons, New York, **1980**
7. Berry, G.C.; Fox, T.G. *Adv. Polym. Sci.* **1968**, *5*, 261.
8. Porter, R.S.; Johnson, J.F. *Chem. Rev.* **1966**, *66*, 1.
9. King, J.A.Jr. *Trends in Macromol. Res.* **1994**, *1*, 1.
10. Sakashita, T.; Shimoda, T. U.S. Patent 5,026,817 **1991**
11. General DOE: G.E.P. Box, W. G. Hunter, and J.S. Hunter *Statistics for Experimenters* John Wiley & Sons, New York, New York 1978.
12. Multiple Regression: *RS/Explore Mulreg Reference Manual* BBN Software Products, Cambridge, MA, 1992; and N. Draper and H. Smith *Applied Regression Analysis, 2nd Ed.* John Wiley & Sons, New York, New York,1981.
13. A number of my colleagues in the polyester industry use the occurrence of this "foaming" transition as an indication of whether their polymerization is working properly or not. However, the molecular significance of this event is not commonly known.
14. The melt strength of the polymeric melt medium is assumed to be proportional to its melt viscosity.
15. All predictions made in this work are given with a confidence range that encompasses a 95% confidence interval. This interval range is typical for experimental work.
16. The traditional (primary) stirrer was used was simple ¼" thick right-handed helixing nickel stirrer shown in pictures 1-8. A custom designed high efficiency stirrer was built with the assistance of Mahari Tjahjadi which consisted of a solid nickel 0.25" x 7.25" oval plate twisted by 90^{0} into a left-handed helix. It contained 6 randomly dispersed 0.25" diameter holes for better turbulent mixing. This latter stirrer allowed a more accurate programming response for phenol mass loss versus time than the original stirrer. Although the on-set foaming times were substantially shorter with the efficiency stirrer, at a given temperature and pressure, the M_w and IV's were the same for both stirrers. Stirring rates were also varied between 200-500 rpm with no noticeable effect.
17. In Analysis of Variance (ANOVA), the F-ratio is a measure of the signal/noise ratio. With sample sets of this reported size, F-ratios of 2 or less indicate that it is unlikely that a real effect is present; F-ratios of 4 or more indicate that it is quite

likely (95% probability) that a real effect is present. F-ratios above 10 indicate a 99+% probability that a real effect is present.

18. Random effects include those related to the experimental set-up and its execution: *i.e.* temperature and pressure fluctuations, weighing errors, variation in sampling points, operator errors, measurement errors, *etc.*.

19. This does not mean that no increase in the melt viscosity has occurred merely that the course monitoring system doesn't register any distinction between 250-600 cps under our reaction conditions (240-300°C; over ~2-3 h period) where on-set occurs. The melt viscosity rises rapidly (20-200 poise/min) after the foaming begins.

BULK FREE RADICAL POLYMERIZATION

Chapter 5

Solvent-Free Stable Free Radical Polymerization: Understanding and Applications

Peter G. Odell, Nancy A. Listigovers, Marion H. Quinlan, and Michael K. Georges

Xerox Research Centre of Canada, Mississauga, Ontario L5K 2L1, Canada

The kinetics of Stable Free Radical Polymerization permits the polymerization to be conducted without solvents. The use of nitroxide as a reversible chain terminator provides control over the polymer architecture in a free radical polymerization. The concentration of the unbound nitroxide capping agent controls the kinetics. The difference in the nitroxide concentration behaviour between styrene and acrylates has been studied over the course of the polymerization. A variety of block copolymers have been synthesized and new GPC techniques developed to elucidate their structure. The polymerizations have also been conducted in supercritical carbon dioxide.

Stable Free Radical Polymerization (SFRP) offers the advantages of free radical polymerization, while sharing many of the attributes of other living polymerization systems, such as the ability to synthesize block, dendrimeric, and hyperbranched copolymers. Unlike other "living" polymerizations, such as anionic or group transfer polymerization (GTP), SFRP is inexpensive, robust, can be performed without any prior purification of monomers, and is compatible with a wide variety of functional groups. In addition, GTP is restricted by its applicability to (meth)acrylates alone. Anionic polymerization, although applicable to both styrenics and (meth)acrylates is limited to homo and block copolymers of finite architecture, since cross-initiation between disparate monomers is not possible. In SFRP, cross-initiation is possible, which substantially broadens its synthetic use.

The key to SFRP is the use of nitroxide stable free radicals to reversibly terminate the propagating chains, thus substantially reducing premature irreversible termination, and enabling monomer to continue to add in a controlled fashion over the course of the polymerization. The replacement of conventional, irreversible termination with reversible termination by a small molecule in the polymerization eliminates the prospect of an autoacceleration, or Trommsdorf effect (*1*). This allows the reaction to be readily carried out in a bulk or solventless manner. Because solventless polymerizations are easy to perform, and do not necessitate the eventual removal of solvent, they were used extensively in the development of the SFRP process, and continue to be used for the synthesis of homopolymers and block copolymers. This chapter examines the chemistry of the SFRP process, under solventless conditions, along with recent developments in the synthesis and characterization of block copolymers. Preliminary results of the SFRP process under supercritical CO_2 polymerization conditions, an alternative route to freeing a polymerization of conventional solvents, are also presented.

All living free radical polymerizations employ a small molecule to reversibly terminate the propagating polymer chain. The first example of this was reported by Borsig et al. who showed that methyl methacrylate oligomers, terminated with primary radicals derived from 1,1,2,2-tetraphenyl-1,2-diphenoxyethane, could initiate a conventional polymerization of methyl methacrylate by the homolytic cleavage of the terminal group from the oligomer chain (*2*). Otsu et al subsequently introduced the use of iniferters to provide a more controlled polymerization (*3-4*). Dithiocarbamate radicals were shown to reversibly terminate growing polymer chains such that the polymer chains increased in molecular weight in a linear fashion with conversion. However, the propensity of the dithiocarbamate radical to initiate new chains during the course of the polymerization, or to lose CS_2, prevented the polymerization from behaving in a well-defined, living manner (*5*). In the 1970's, a series of papers were published showing that alkoxyamines were thermally unstable and would regenerate the nitroxide, or a hydroxylamine, upon continued heating (*6-7*). In 1984, Rizzardo and Solomon successfully used the 1-(2-cyano-2'-propoxy)-2,2,6,6-tetramethylpiperidine adduct to initiate the synthesis of acrylate oligomer, although in low yields (*8-9*). Georges et al built on this initial result to synthesize polystyrene resins with polydispersities narrower than what was considered theoretically possible (<1.5) by free radical polymerization (*10*). This discovery stimulated new and renewed interest in the concept of a living free radical polymerization, and lead to an exponential increase in the number of papers published in the field. In 1995, Sawamoto reported another living radical polymerization system using a halogen as the terminating species and a ruthenium/aluminum complex to reversibly remove it (*11*). In a similar manner, and in a process referred to as atom transfer radical polymerization (ATRP), Matyjaszewski utilizes a halogen in combination with a copper complex to reversibly terminate polymer chains (*12*). Other workers who have made significant contributions to the application and understanding of nitroxide moderated polymerization include Veregin, (*13-17*) Hawker, (*18-26*) Matyjaszewski, (*27-30*) Moad, (*31-32*) Fukuda, (*33-39*) Yoshida, (*40-41*) and Priddy (*42-43*).

Mechanism

The following section explores the basic mechanism of the polymerization in three parts: initiation, propagation, and termination. Elucidation of all the features of the polymerization kinetics is ongoing, but significant progress has been made to date (*13-17,22-32,34-39*).

Initiation. Initiation requires the creation of a carbon centered radical, in the presence of the oxygen centered radical of the nitroxide, and only the former radical is capable of adding monomer. Two general approaches have been employed: one based on the use of conventional initiators (Scheme 1) and a second, employing specially synthesized adducts that homolytically cleave to provide the required species (Scheme 2). When conventional initiators such as benzoyl peroxide or AIBN are employed, one

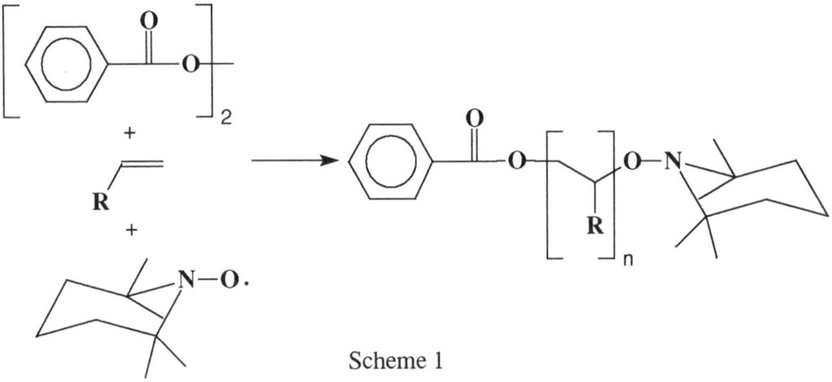

Scheme 1

starts with a large excess of nitroxide which captures the nascent radicals. The amount of nitroxide used is determined by the efficiency of the initiator. The principle of the persistent radical effect mandates that sufficient excess nitroxide remains to prevent the irreversible coupling of two carbon centred radicals (*44-46*). In styrene polymerizations, the excess concentration of nitroxide over radical chain ends is about three orders of magnitude (10^{-5} M vs 10^{-8} M). As the length of the radical capped chain grows its mobility declines, as does the requirement for the relatively high nitroxide concentration. When the adducts of Scheme 2 are employed as initiators, in principle there is no excess nitroxide, but the same requirement for it exists, and some of the carbon centered radical fragments must inevitably couple to provide the build-up of the nitroxide concentration. A variation of both types of initiation is to use nitroxide capped oligomers produced via conventional initiation, and subsequently isolated for use in a later polymerization.

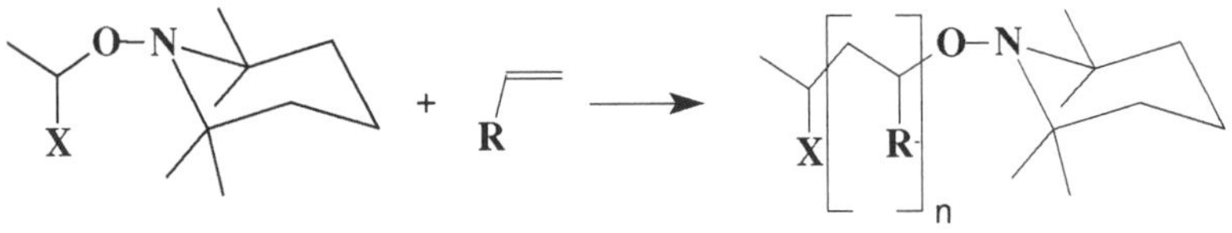

Scheme 2

Propagation. The rate of polymerization of the SFRP process is moderated by reversibly capping the propagating chain with the nitroxide as expressed by equation 1:

$$P_n^\bullet + T^\bullet \underset{\longleftarrow}{\overset{K_L}{\longrightarrow}} L_n \qquad (1)$$

where $P_n^\bullet$ is the uncapped, free radical terminated polymer chain, $T^\bullet$ is the nitroxide, and L_n is the reversibly capped polymer chain. The concentration of $T^\bullet$ can be readily monitored by ESR during the course of a polymerization. With this knowledge in hand, $[P_n^\bullet]$ can be derived from equilibrium (1) to give:

$$[P_n^\bullet] = L_n/K_L[T^\bullet] \qquad (2)$$

This expression can now be substituted into the usual differential equation for the rate of propagation (equation 3) to yield equation 4, the differential rate equation for propagation via SFRP.

$$\frac{d[M]}{dt} = -k_p[M]\Sigma[P_n^\bullet] \qquad (3)$$

$$R_p = k_p [P^\bullet][M] = k_p \frac{[L_n]}{K[T^\bullet]}[M] \qquad (4)$$

The impact of the nitroxide concentration on the polymerization of styrene at 125°C, initiated with BPO at different TEMPO concentrations, with and without the additive, 2-fluoro-1-methylpyridinium p-toluenesulfonate (FMPTS), is shown qualitatively in Figure 1. To the left lie four curves describing the logarithmic conversion against time, and on the right, the corresponding change in nitroxide concentration over the course of the polymerization.

It can be readily seen in Figure 1 that the observed rate of polymerization can be increased by reducing the concentration of the stable free radical nitroxide. As $T^\bullet$ is removed from the reaction medium via reaction with FMPTS, the equilibrium in Equation 1 is driven to the left, increasing the concentration of reactive polymer chains, and thus the rate of polymerization. Once the nitroxide levels have been reduced, the polymerization system can control these levels through a combination of autoinitiation and termination. The generation of new chains by autoinitiation via the Mayo mechanism, helps reduce the nitroxide level if it builds up beyond what is required for rapid polymerization. Conversely, if the polymerization is disturbed and too little nitroxide is present, the mobile radicals produced by autopolymerization offer a termination route for $L_n^\bullet$ before a rise in the ratio $[L_n^\bullet]/[T^\bullet]$ can result in an autoacceleration.

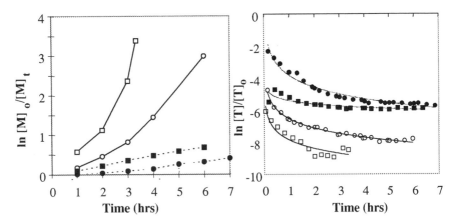

Figure 1: A: Observed rate of conversion vs time. B: Nitroxide concentration vs time for the same polymerization. ─□─ 1:1.1 BPO:TEMPO; ─○─ 1:1.3 BPO:TEMPO; ─■─ 1:1.1 BPO:TEMPO w. FMPTS; ─●─ 1:1.3 BPO:TEMPO w. FMPTS;

A time dependent relationship for the nitroxide concentration can be derived from the above curves of Figure 1B, in the form of equation 5. The values of constants b and S are determined empirically from the curves of Figure 1B.

$$[T] = [T]_o bt^S \qquad (5)$$

In a well-behaved system, the number of living chains, $[L_n]$ will be essentially constant and approximately equal to the initial concentration of the nitroxide. With this constraint, equation 4 can be rewritten as an integral in equation 6.

$$\int \frac{d[M]}{[M]} = -k \int \frac{dt}{[T]_o bt^S} \qquad (6)$$

In these polymerizations S>0, and integration provides equation 7.

$$\ln\left\{\frac{[M]_0}{[M]_t}\right\} = kct^{(1-S)} \qquad (7)$$

where c is given by equation 8.

$$c = \left\{\frac{1}{b(1-S)[T]_o}\right\} \qquad (8)$$

Equations 7 and 8 can be used to obtain a logarithmic plot of conversion against a time axis that is corrected for variance in nitroxide concentration ($\ln([M]_0/[M]_t)$ versus $ct^{(s+1)}$). Simply by accounting for the change in free nitroxide concentration, the rates of the four polymerizations of Figure 1A become nearly coincident.

Acrylate Polymerization. Extending the bulk SFRP process from styrene to alkyl acrylate synthesis proved to be more challenging than expected. Initial syntheses of random copolymers of styrene with low amounts of n-butyl acrylate were straightforward providing high yields of narrow polydispersity resins. However, as the acrylate content was increased, the polymerizations became more difficult. Higher polydispersities and lower conversions resulted with increasing levels of acrylate (Table 1). Initial attempts at homopolymers of n-butyl acrylate were unsuccessful resulting in low molecular weight resins with conversions of less than 5%. Attempts at chain extensions of polystyrene with n-butyl acrylate also proved difficult, proceeding slowly to low molecular weight blocks and then stopping.

Table 1: Influence of increasing acrylate content in a SFR random copolymerization with styrene.

Mole% n-BuAc	$M_w(10^{-3})$	$M_n(10^{-3})$	PD	Conv. %
24.7	32.5	24.6	1.32	89.3
43.5	38.2	24.8	1.54	83.0
64.1	36.4	24.4	1.49	65.2
81.2	10.6	6.1	1.73	33.4

ESR experiments showed that, in contrast to styrene polymerizations, in which excess nitroxide could be maintained at low levels, the free nitroxide levels in acrylate polymerizations increase with time (Figure 2), eventually reaching levels that inhibit the polymerization.

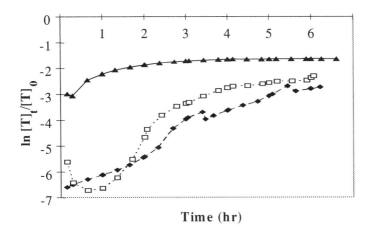

Time (hr)

Figure 2: Increase in free nitroxide concentration during polymerization with ethylhexylacrylate (◆ 40% with styrene, □ 75% with styrene, ▲ 100% acrylate)

The major obstacle to acrylate polymerization by the SFRP process is this increase of the free nitroxide concentration. This is in contrast with the polymerization of styrene, where a wide range of nitroxide levels can be tolerated resulting only in a variation of observed rates. A broadening of the molecular weight distribution is only observed at very low levels of nitroxide in styrene. It has been shown for styrene that at low nitroxide levels, a portion of conventional free radical polymerization occurs until the nitroxide level rises (47). In acrylates the picture is not as clear. The rate constants of propagation for acrylates are greater than in styrene. This implies that a deviation from ideal SFRP conditions may result in more profound changes in the polymerization. While the potential for acrylate autopolymerization is not understood to the extent that it is in styrene, long term heating of acrylate monomer with small amounts of nitroxide in the absence of initiator provides no detectable polymer. Anecdotal reports from the acrylate industry suggest that sudden spontaneous polymerization as a result of impurities are known. Reports of self-initiated thermal polymerization are rarer.

To advance the acrylate polymerization, control of the nitroxide level was required. The impact of changes in the nitroxide level was examined by controlled addition of small amounts of camphorsulfonic acid (CSA) (48) or (FMPTS) (49-50) to acrylate polymerizations that were continuously monitored by ESR. The polymerizations in this study were initiated by a nitroxide capped polystyrene oligomer. The molecular weight distributions and the observed rates in the bulk polymerization of n-butylacrylate varied considerably as the nitroxide level was manipulated.

In the first of a series of experiments, rapid polymerization from the starting TEMPO capped polystyrene oligomer occurred initially since the starting nitroxide concentration was very low. As the nitroxide concentration rose, the polymerization slowed. It remained well-behaved to the extent that the molecular weight distribution did not broaden. However, the rate of polymerization failed to increase even though the nitroxide concentration, $[T^\bullet]$, was driven back down to 10^{-4} M. The initial rapid polymerization occurred at a concentration below this. Thus, at nitroxide concentrations of 10^{-4} M and above, SFRP was very slow; chain growth was uniform or suppressed entirely.

Alternatively, additional nitroxide was added to mimic the initial rise in concentration experienced in the previous experiment. The nitroxide was then driven to lower levels with repeated additions of FMPTS. In the first 30 minutes, a small but uniform growth in all the chains was observed. As the nitroxide level dipped below 10^{-4} M, molecular weight growth increased, but the dispersity broadened. Significantly, as the nitroxide level once again rose at about 200 minutes into the reaction, molecular weight growth virtually ceased. At 270 minutes as $[T^\bullet]$ was lowered again and approached 5×10^{-6} M, molecular weight growth proceeded with renewed vigour. There was evidence in the GPC traces of low molecular weight chains that have not continued to grow. This data indicates that nitroxide concentrations of between 5×10^{-5} and 5×10^{-6} M provide a good rate of polymerization, but at a cost in the narrowness of the polydispersity.

Further reduction in the concentration of the free nitroxide during the course of the polymerization provided even higher rates of molecular weight growth. However, as the nitroxide level was driven to about 5×10^{-7} M, the polymerization proceeded at a dramatic rate and achieved molecular weights in excess of 100k. Elsewhere in this chapter the nature of such rapidly produced polymers is described in detail. We have found that driving the nitroxide concentration to very low levels can yield high molecular weight polymers that are capable of further growth.

In summary, the role of the nitroxide concentration remains pivotal in SFRP of acrylates, but the behaviour is more complex than with styrenic monomers.

Termination. In a perfectly living polymerization, termination does not occur. A *living radical* polymerization, of which SFRP is a prominent example, does however experience a variety of possible termination events. The degree and significance of these events determines the quality of the polymer product. In a living radical polymerization, termination (Term) consists of two parts:

$$\text{Term} = \text{Term}_{rev} + \text{Term}_{irr}$$

where the reversible termination is the desired outcome and the irreversible termination must be minimized.

Any event that leads to a decrease in the number of radical chain ends [$P_n^\bullet$] will result in a rise in [$T^\bullet$] and a reduction in the rate of polymerization. In addition to additives that destroy nitroxide, the autopolymerization of styrene can consume "excess" nitroxide (*30, 33-34*). Styrene autopolymerization likely contributes to the robust, self-correcting control over the nitroxide level during the polymerization of this monomer. The differences in the ease of application to various classes of vinyl monomers may arise from the differences in termination, with the principles of propagation being the same. The "quality" of the polymer is determined by the extent and type of termination reactions, as it is in conventional free radical polymerization. The techniques required to quantitate this quality are discussed later in this chapter.

Block Copolymer Synthesis

Diblock copolymers were obtained by a stable free radical polymerization, in a two step process, with each step consisting of a bulk polymerization (*51*). The first block was synthesized by a "living" polymerization, isolated with retention of the TEMPO end group, and used as a macroinitiator in the second step. The nitroxide terminated homopolymer was dissolved in a second monomer and heated to a temperature at which dissociation of the nitroxide end group is facile. This resulted in chain extension of the first block, via a "living" polymerization, to form the second block. In theory, either block could be synthesized first, but in practice, it was often more straightforward to carry out the synthesis in a particular order which varied depending on the diblock in question. For example, in a polystyrene-*b*-poly(4-vinylpyridine) copolymer, the poly(4-vinylpyridine) block was synthesized first, with a relatively high nitroxide concentration, followed by chain extension with styrene monomer. Chain extension or generation of the second block is usually carried out under low nitroxide conditions which are unsuitable for 4-vinylpyridine due to its propensity to autopolymerize.

The range of block copolymers that have been made is illustrated in Table 2. Polystyrene-*block*-poly(alkyl acrylate) copolymers are of particular interest due to their potential application as surface active agents, pigment dispersants, flocculants, and compatibilizers for polymer blends. Through careful manipulation and control of the nitroxide level, polystyrene-*b*-poly(*n*-butylacrylate) copolymers ranging from 5K to 400K in molecular weight with compositions ranging from 2mol% to 90mol% polystyrene have been synthesized by this process. In this case, either block can be synthe-

Table 2: Composition & MW ranges for block copolymers produced with SFRP

A-block Monomer	B-block Monomer	mol% A	mol% B	Mw(x10⁻
styrene	*n*-butyl acrylate	2-95	98-5	5-400
	isoprene	60	40	50
	t-butyl acrylate	5-95	95-5	3-100
	chloromethylstyrene	10-90	90-10	10-50
	4-vinylpyridine	70-95	30-5	10-20
	4-vinylbenzoic acid	16	84	16
	butadiene	50-80	50-20	5-20
	hydroxymethylstyrene	80-95	20-5	10-25
	4-acetoxystyrene	30-95	70-5	4-50
n-butyl acrylate	*t*-butyl acrylate	40-60	60-40	14-20

sized first, demonstrating that cross-initiation of these two monomers is possible. Substitution of *n*-butyl acrylate with *t*-butyl acrylate results in a comparable set of diblock copolymers which can be hydrolyzed to give polystyrene/polyacrylic acid block copolymers. These amphiphilic block copolymers have application as pigment dispersants, and can be tailored for water, or organic solubility depending on the relative length of the two blocks. For example, copolymers with short PAA blocks and longer PS blocks are excellent dispersants for a proprietary Xerox pigment in toluene, whereas copolymers with short polystyrene blocks and long PAA blocks have been used to disperse carbon black in aqueous media.

The synthesis of styrene/diene diblock and triblock copolymers has also proved feasible by the SFRP process. For example, TEMPO-terminated polystyrene reacts with isoprene at 135°C to produce a diblock copolymer. The diblock copolymer can subsequently be isolated, purified and redissolved in styrene to form a triblock of polystyrene-*b*-polyisoprene-*b*-polystyrene. The GPC chromatogram (Figure 3) shows the step-wise shift to higher molecular weight as each block is added.

In the examples discussed above, styrene was used in combination with non-styrenic monomers. Styrene has also been used in combination with functionalized styrenics to yield amphiphilic diblock copolymers. SFRP is readily applied to monomers such as 4-acetoxystyrene and chloromethylstyrene. In contrast to anionic polymerization, the acetoxy and chloromethyl groups remain intact without protection. The allure of these functional groups is that they are readily modified which allows access to many complex structures in a minimum number of steps (see Scheme 3).

For example, poly(4-hydroxystyrene) which has been used as a photoresist is easily obtained from poly(4-acetoxystyrene) via a base hydrolysis using ammonium hydroxide. Poly(chloromethylstyrene) is a very versatile material due to the facile displacement of the benzylic chloride. It can serve as a site for cross-linking or as illustrated in Scheme 3, can be functionalized to give an amphiphilic block copolymer.

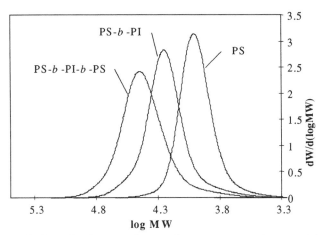

Figure 3: Growth of styrene homopolymer to styrene-isoprene block copolymer and then to polystyrene-isoprene-polystyrene triblock.

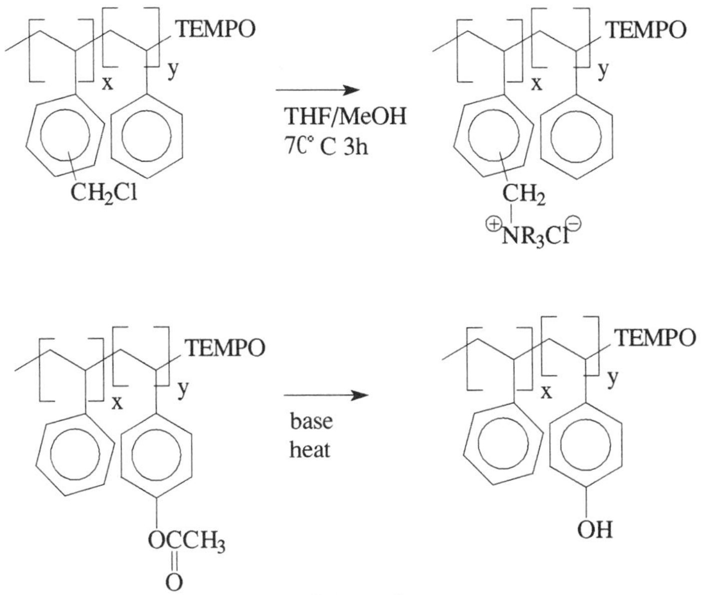

Scheme 3

In summary, many vinyl monomers capable of conventional free radical polymerization have been incorporated into di- and triblock copolymers via the SFRP process. This has resulted in a diverse range of materials in terms of both composition and molecular weight.

Characterization of Polystyrene-*b*-Poly(acrylate) Copolymers via GPC Techniques

Since irreversible termination reactions can be present in varying degrees, depending on the polymerization conditions and nature of the monomers, detection and quantitation of the products of these side reactions is vital to the optimization of the synthesis. Conventional gel permeation chromatography (GPC) characterization of the diblock copolymers using refractive index (RI) detection proved insufficient to determine the presence of homopolymer contaminants. For example, Figure 4 shows a GPC comparison plot of a polystyrene-*b*-poly(*n*-butyl acrylate) copolymer versus its TEMPO-terminated polystyrene precursor. There is no readily discernable peak attributable to the 10 K polystyrene starting material in the block copolymer.

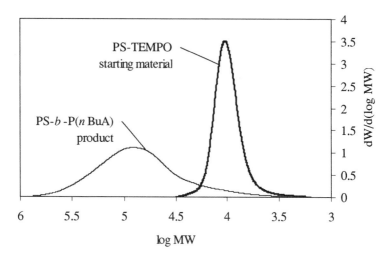

Figure 4: Comparison of GPC chromatograms of PS-TEMPO starting material and PS-*b*-P(*n*BuA) copolymer derived from it.

If RI detection, which is sensitive to both polymers, is combined with UV detection at 254 nm, which is sensitive only to polystyrene, a different picture emerges (Figure 5). A very distinct peak attributed to the polystyrene starting material was observed in the UV chromatogram. Further optimization of the reaction conditions has enabled the synthesis of polystyrene-*b*-poly(*n*-butyl acrylate) copolymers in which the

RI and UV peaks are congruent, which is indicative of a homogeneous block copolymer.

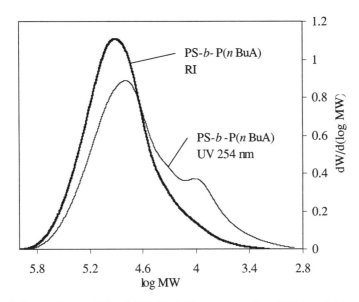

Figure 5: Comparison of RI and UV GPC Chromatographs of PS-*b*-poly(*n*BuA) copolymer revealing contamination with starting material

In cases where the first block of the copolymer is an acrylate, the RI/UV detector comparison is less definitive for the presence of unreacted starting material. In that case, a GPC technique using *o*-xylene as the eluent allows the detection of the starting acrylate. *Ortho*-xylene is a high refractive index solvent in which poly(butyl acrylate) gives a negative RI signal, while PS maintains a positive RI signal. In the example shown in Figure 6, the starting material has a peak maximum (Mp) of 51K and a negative signal. Chain extension with polystyrene leads to a block copolymer with an Mp=146K and a positive RI signal. There is no indication of a negative peak or precipitous drop in the RI signal at the low molecular weight end of the chromatogram, strongly suggesting all the starting material has reacted.

Chromatographic separation techniques combining HPLC and GPC have been used to quantify the amount of homopolymer (if any) present, but are beyond the scope of this chapter.

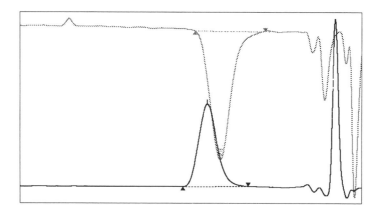

Figure 6: GPC chromatograms of P(n-BuA)-TEMPO starting
material and the resulting P(n-BuA)-b-PS-TEMPO copolymer

SFRP Supercritical CO_2 Polymerization

The application of supercritical carbon dioxide to the SFRP process has been investigated as a solventless route to block copolymers (52). Supercritical CO_2 provides an extracting medium for residual monomer from the first block formed. In addition, the plasticization of the polymer, coupled with the increased diffusivity of monomer dissolved in the supercritical fluid, can aid the rate and extent of growth of the second block. Thus a single pot synthesis of block copolymers becomes a possibility.

The initial investigation of stable free radical polymerization (SFRP) in supercritical carbon dioxide presented two key findings. First, below about 30 w/v% styrene in the reactor, the polymer stopped growing at between 2000 to 4000 Daltons, while above this loading the polymerization performed similarly, though somewhat more slowly, to a conventional SFRP styrene reaction. McArthy (53) has demonstrated that even in a dissimilar polymer matrix, monomer can migrate to a radical site and polymerize. And DeSimone has explored the nature of free radical initiation (54) and propagation (55) in supercritical CO_2 and concluded that it is a suitable polymerization medium. Thus, this cessation of polymerization is not a feature of a radical polymerization in supercritical CO_2, rather it relates specifically to SFRP. The arrested polymerization phenomena at low styrene (CO_2 rich) loadings could be overcome with the addition of small amounts of sulfur dioxide to the polymerization. It was demonstrated that the SO_2 consumed the nitroxide, in a similar fashion to CSA or FMPTS. And this represented another example of the nitroxide concentration controlling the outcome of the SFRP process.

Conclusion

Stable Free Radical Polymerization lends itself well to solventless polymerization. SFRP is most effective with styrenic monomers, but recent advances have allowed the practical polymerization of acrylic monomers. Most of the mechanistic features of the polymerization can be explained in terms of the concentration of the free nitroxide. Controlling this concentration is the key to a successful polymerization. With the improvement in the acrylate polymerization, numerous block copolymers have been synthesized, many of these with sufficient purity to be used in demanding electro-optical applications. New characterization techniques were developed in order to quantitate the quality of the block copolymers. These techniques are required since conventional chain termination has been minimized but not eliminated in these polymerization systems. Lastly, supercritical CO_2 offers further potential for providing complex macromolecular structures in the absence of organic solvents.

Literature Cited:

1. Saban, M.; Georges, M. K.; Veregin, R. P. N.; Hamer, G. K.; Kazmaier, P. M. *Macromolecules* **1995**, *28*, 7032.
2. Borsig, E.; Lazár, M.; Capla, M; Flouan, S. *Angew. Makromol. Chem.* **1969**, *9*, 89.
3. Otsu, T.; Yoshida, M.; Tazaki, T. *Makromol. Chem., Rapid Commun.* **1982**, *3*, 133.
4. Endo, K.; Murata, K.; Otsu, T. *Macromolecules* **1992**, *25*, 5554.
5. Turner, R. S.; Blevins, R. W. *Macromolecules* **1990**, *23*, 1856.
6. Bolsman, T. A. B. M.; Blok, A. P.; Frijns, J. H. G. *J. R. Neth. Chem. Soc.* **1978**, *97*, 313.
7. Grattan, D. W.; Carlsson, D. J.; Howard, J. A.; Wiles, D. M. *Can. J. Chem.* **1979**, *57*, 2834 and references therein.
8. Solomon, D. H.; Rizzardo, E.; Cacioli, P. *European* Patent, **1985**, 135280.
9. Rizzardo, E, *Chem, Aust.* **1987**, *54*, 32.
10. Georges, M. K.; Veregin, R. P.N.; Kazmaier, P. M.; Hamer, G. K. *Macromolecules* **1993**, *26*, 2987.
11. Kato, M.; Kamigaito, M.; Sawamoto, M.; Higashimura, T. *Macromolecules* **1995**, *28* 1721.
12. Wang, J.; Matyjaszewski, K. *J. Am. Chem. Soc.* **1995**, *117*, 5614
13. Veregin, R.P.N.; Georges, M. K.; Kazmaier, P. M.; Hamer, G. K. *Macromolecules* **1993**, *26*, 5316.
14. Veregin, R.P.N.; Georges, M. K.; Hamer, G.K.; K. Kazmaier, P. M. *Macromolecules* **1995**, *28*, 4391.
15. Veregin, R. P. N.; Odell, P. G.; Michalak, L. M.; Georges, M. K. *Macromolecules* **1996**, *29*, 2746.
16. Veregin, R. P. N.; Odell, P. G.; Michalak, L. M.; Georges, M. K. *Macromolecules* **1996**, *29*, 3346.
17. Veregin, R.P.N.; Kazmaier, P. M.; Odell, P.G.; Georges, M. K. *Chem. Lett.* **1997**, 467.

94

18. Hawker, C. J. *Angew. Chem., Int. Ed. Engl.* **1995**, *34*, 1456.
19. Hawker, C. J.; Frechet, J. M. J.; Grubbs, R. B.; Dao, J. *J. Am. Chem. Soc.* **1995**, *117*, 10763.
20. Leduc, M. R.; Hawker, C. J.; Dao, J.; Frechet, J. M. J. *J. Am. Chem. Soc.* **1996**, *118*, 11111.
21. Hawker, C. J.; Mecerreyes, D.; Ecle, E.; Dao, J.; Hedrick, J. L.; Barakat, P.; Dudois, P.; Jerome, R.; Volksen, W. *Macromol. Chem. Phys.* **1997**, *198*, 155.
22. Hawker, C. J. *J. Am. Chem. Soc.* **1994**, *116*, 11185.
23. Hawker, C. J.; Barclay, G. G.; Orellana, A; Dao, J.; Devonport, W. *Macromolecules* **1996**, *29*, 5245.
24. Hawker, C. J.; Barclay, G. G.; Dao, J. *J. Am. Chem. Soc.* **1996**, *118*, 11467.
25. Devonport, W.; Michalak, L.; Malmstrom, E.; Mate, M.; Kurdi, B.; Hawker, C. J.; Barclay, G. G.; Sinta, R. *Macromolecules* **1997**, *30*, 1929.
26. Michalak, L.; Malmstrom, E.; Devonport, W.; Mate, M.; Hawker, C. J. *Polym. Prepr., Am. Chem. Soc., Div. Polym. Chem.* **1997**, *38 (1)*, 727.
27. Greszta, D.; Mardare, D.; Matyjaszewski, K. *Macromolecules* **1994**, *27*, 638.
28. Matyjaszewski, K.; Gaynor, S.; Greszta, D.; Mardare, D.; Shigemoto, T. *Macromol. Symp.* **1995**, *98, 73*.
29. Matyjaszewski, K.; Greszta, D. *Macromolecules* **1996**, *29*, 5239.
30. Greszta, D.; Matyjaszewski, K. *Macromolecules* **1996**, *29*, 7661.
31. Moad, G.; Rizzardo, E. *Macromolecules* **1995** *28*, 8722.
32. Moad, G.; Ercole, F.; Krstina, J.; Moad, C. L.; Rizzardo, E.; Thang, S. H. *Polym. Prepr., Am. Chem. Soc., Div. Polym. Chem.* **1997**, *38 (1)*, 744.
33. Fukuda, T.; Terauchi, T.; Goto, A.; Ohno, K.; Tsujii, Y.; Miyamoto, T.; Shimizu, Y. *Macromolecules* **1996** *29*, 3050.
34. Fukuda, T.; Terauchi, T. *Chem. Lett.* **1996**, 293.
35. Fukuda, T.; Terauchi, T.; Goto, A.; Ohno, K.; Tsujii, Y.; Miyamoto, T.; Kobatkae, S.; Yamada, B. *Macromolecules* **1996** *29*, 6393.
36. Goto, A.; Terauchi, T.; Fukuda, T.; Miyamoto, T. *Macromol. Rapid Commun.* **1997** *18*, 673.
37. Fukuda, T.; Goto, A. *Macromol. Rapid Commun.* **1997** *18*, 683.
38. Goto, A.; Fukuda, T. *Macromolecules* **1997** *30*, 4272.
39. Goto, A.; Fukuda, T. *Macromolecules* **1997** *30*, 5183.
40. Yoshida, E.; Ishizone, T.; Hirano, A.; Nakahama, S.; Takata, T.; Endo, T. *Macromolecules*, **1994**, *27*, 3119.
41. Yoshida, E.; Sugita, A. *Macromolecules*, **1996**, *29*, 6422.
42. Howell, B. A.; Priddy, D. B.; Li, I.Q.; Smith, P. B.; Kastl, P. E. *Polymer Bulletin*, **1996**, *37*, 451.
43. Zhu, Y.; Howell, B. A.; Priddy, D. B. *Polym. Prepr., Am. Chem. Soc., Div. Polym. Chem.* **1997**, *38 (1)*, 97.
44. Baldovi, M. V.; Mohtat, N.; Scaiano, J. C. *Macromolecules* **1996** *29*, 5497.
45. Fischer, H. *J. Am. Chem. Soc.* **1986**, *108*, 3925.
46. Fischer, H. *Macromolecules* **1997** *30*, 5666.
47. MacLeod, P. J.; Veregin, R. P. N.; Odell, P.G.; Georges, M. K. *Macromolecules* **1997**, *30*, 2207.

48. Veregin, R. P. N.; Odell, P.G.; Michalak, L. M.; Georges, M. K. *Macromolecules* **1996**, *29*, 4161.
49. Odell, P.G.; Veregin, R. P. N.; Michalak, L. M.; Brousmiche, D.; Georges, M. K. *Macromolecules* **1995**, *28*, 8453.
50. Odell, P.G.; Veregin, R. P. N.; Michalak, L. M.; Georges, M. K. *Macromolecules* **1997**, *30*, 2232.
51. Listigovers, N. A.; Georges, M. K.; Odell, P.G.; Keoshkerian B. *Macromolecules* **1996**, *29*, 8992.
52. Odell, P. G.; Hamer, G. K. *Polym. Mater. Sci. Eng.* **1996**, *74*, 404.
53. Watkins, J. J.; McCarthy, T. J. *Macromolecules* **1994**, *27*, 4845.
54. Guan, Z.; Combes, J. R.; Menceloglu, Y. Z.; DeSimone, J. M. *Macromolecules* **1993**, *26*, 2663.
55. van Herk, A. M.; Manders, B. G.; Canelas, D. A.; Quadir, M. A.: DeSimone, J. M. *Macromolecules* **1997**, *30*, 4780.

Chapter 6

Bulk Atom Transfer Radical Polymerization

Krzysztof Matyjaszewski

Department of Chemistry, Carnegie Mellon University, 4400 Fifth Avenue, Pittsburgh, PA 15213

Bulk controlled radical polymerizations allow for the preparation of polymers with controlled molecular weights, end functionalities and low polydispersities. It is possible, using these systems, to control chain architecture and composition and to prepare segmented copolymers such as block and graft copolymers. Atom transfer radical polymerization (ATRP) is one of the most robust controlled radical systems and has been successfully used for bulk polymerization of substituted styrenes and functional (meth)acrylates. Several peculiarities of bulk controlled radical polymerization such as the absence of the Trommsdorf effect, chain length dependent termination, importance of side reactions and contribution of self-initiation are discussed.

Controlled/"living" radical polymerization enables preparation of well-defined and complex macromolecular architectures for a larger number of monomers and under less stringent conditions than those used for ionic processes. Both low and high molecular weight polymers with well-defined molecular weights, low polydispersities and novel functionalities have been synthesized using various controlled radical polymerization techniques. All of these techniques rely on the suppression of the termination reaction between growing radicals by using a low stationary concentration of radicals via dynamic equilibration between free radicals and various types of dormant species. In addition, in controlled radical polymerization initiation is fast and completed at low monomer conversion. One of the big advantages of controlled radical polymerization is the absence of the Trommsdorf (gel) effect which enables one to carry out polymerization without any solvent. This paper is focused on a discussion of the peculiarities of bulk controlled radical polymerization, especially atom transfer radical polymerization (ATRP)

Fundamentals of Controlled Radical Polymerizations

Chain breaking reactions such as transfer and termination should be absent in truly living polymerizations.(1) Thus, since growing radicals always terminate, radical polymerization can never be truly living but under appropriate conditions can be controlled and exhibit many characteristic features of living polymerization such as preparation of polymers with low polydispersities, linear growth of molecular weight with conversion and good control of end group functionalities. In order to generate such controlled (or "living") polymerization, it is necessary to significantly increase the proportion of non-terminated chains (>90%) and to provide equal probability of growth for all chains by ensuring fast and quantitative initiation.(2) Both of these requirements are very different from the conventional radical polymerization in which initiation is slow and nearly all chains terminate by coupling/disproportionation.(3)

Taking into account typical values of rate constants of propagation ($k_p \approx 10^{3\pm1}$ $M^{-1}s^{-1}$) and termination ($k_t \approx 10^{7\pm1}$ M^{-1} s^{-1}), it is necessary to establish a low concentration of growing radicals ($[P\bullet] \approx 10^{-8\pm1}$ M) in order to reduce the probability of termination. This concentration is similar to that used in conventional processes targeted for high molecular weight polymers ($M_n \geq 100,000$). However, degrees of polymerization in controlled polymerization should be defined by the ratio of concentrations of reacted monomer and growing chains (introduced initiator if initiation is fast), $DP = \Delta[M]/[I]_o$. Thus, to prepare a polymer with $M_n \approx 100,000$ ($DP \approx 1,000$) using bulk conditions ($[M]_o \approx 10$ M), the concentration of the initiator should be in the range of $[I]_o \approx 10^{-2}$ M, which is 10^6 higher than that of free radicals. Because initiator should be rapidly consumed for the controlled polymerization, but it can not generate radicals quantitatively (to avoid termination), a dynamic equilibrium between the majority of dormant chains and the minute amount of free radicals should be established. Under such conditions, termination is less probable, but all macromolecules continuously grow generating well-defined chains, provided that exchange between the dormant and active species is sufficiently fast; especially important is that the rate of deactivation (k_{deact}) is faster than the rate of propagation (k_p):

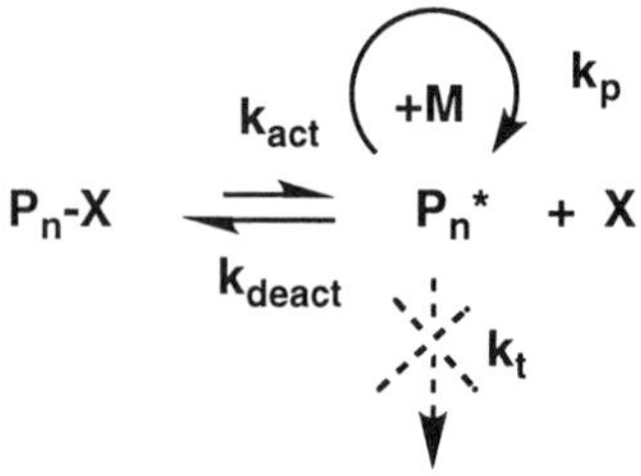

$$P_n\text{-}X \underset{k_{deact}}{\overset{k_{act}}{\rightleftharpoons}} P_n^* + X \quad (+M) \, k_p \quad k_t$$

(1)

There are three general approaches to controlled radical polymerization.*(2)* They include degenerative transfer, reversible formation of persistent radicals and homolytic cleavage of covalent species which include both catalyzed and non-catalyzed reactions.

Degenerative transfer is probably the simplest system because it relies on conventional radical initiators which slowly decompose in the presence of efficient transfer agents. Ideally, transfer agents mimick growing chains in the dormant form and therefore they define the total number of growing chains:

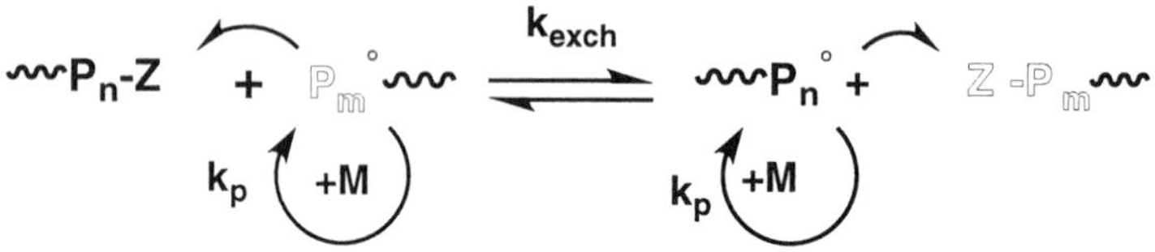

$$\text{\textasciitilde}P_n\text{-}Z + P_m^\circ\text{\textasciitilde} \underset{k_p (+M)}{\overset{k_{exch}}{\rightleftharpoons}} \text{\textasciitilde}P_n^\circ + Z\text{-}P_m\text{\textasciitilde} \quad k_p (+M)$$

(2)

Among the best transfer agents are alkyl iodides (Z=I) which were successfully used in the polymerization of styrene,*(4)* acrylates *(5)* and fluoroalkenes.*(6)* Group Z may also become a non-polymerizable unsaturated group, like in the addition-fragmentation process for methacrylates.*(7)* In all cases, the number of chains is defined by the concentration of R-Z species which should be much higher than that of the decomposed radical initiator ($[RZ]>>[I]_o$). A key feature for control is that the rate constant of exchange should be comparable or faster than the rate constant of propagation, if polymers with low polydispersities are targeted.

The second approach is based on reversible trapping of growing radicals with compounds having an even number of electrons, leading to persistent radicals. Ideally, these radicals should not react directly with alkenes, and should only reverisbly dissociate to provide growing radicals:

$$\text{~}P_n^\circ \; + Z \; \underset{k_{act}}{\overset{k_{deact}}{\rightleftharpoons}} \; \{\text{~}P_n\text{-}Z\}^\circ$$

$$k_p \overset{\curvearrowleft}{(+M)}$$

(3)

It has been proposed that this mechanism may operate in the presence of phosphites,(8) stilbenes, (9) some aluminum (10) and chromium (II) derivatives(11).

The final, third approach seems to be the most successful at present and it is based on the homolytic cleavage of a relatively weak covalent bond in the dormant species. This cleavage can be spontaneous (thermal) or catalytic (e.g. redox process in the presence of transition metal compounds, Y):

$$\text{~}P_n\text{-}X \quad (+\,Y) \quad \underset{k_{deact}}{\overset{k_{act}}{\rightleftharpoons}} \quad \text{~}P_n^\circ \; + X \bullet (Y)$$

$$k_p \overset{\curvearrowleft}{(+M)}$$

(4)

For the non-catalyzed process, relatively stable radicals, X°, should be formed. They should not react directly with monomer but only reversibly scavenge growing radicals. Examples, of such species include nitroxyl radicals,(12-16) dithiocarbamyl radicals,(17) bulky organic radicals (Ar_3C°),(18,19) inorganic/organometallic radicals such as Cr(III) (20) species or Co(II) species with the corresponding ligands (e.g. porphyrines or Schiff bases).(21,22) In some cases, side reactions do occur such as very efficient β-H transfer observed for methacrylates in the presence of Co(II),(23) or slow initiation by dithiocarbamyl radicals, which also decompose to reactive aminyl radicals and CS_2.(24) The non-catalyzed process requires one group X per chain, which may be relatively expensive, especially for shorter polymer chains

The catalyzed process relies on less expensive alkyl halides and pseudohalides which generate radicals by the homolytic cleavage of a C-X bond in the presence of a catalytic amount of the redox-active transition metal compounds, e.g. Cu(I),(25) Ru(II),(26) Fe(II),(27,28) Ni(II) species.(29-31) The halogen atom is reversibly transferred between a growing chain end and transition metal in an inner sphere electron transfer process, or atom transfer process. This process was named atom transfer radical polymerization (ATRP) (25) because atom transfer is the key feature controlling this reaction, and radical intermediates are responsible for the chain growth. In addition, it also correlates with atom transfer radical additions which are well established in organic synthesis.*(32)*

Atom Transfer Radical Polymerization

Probably the most thoroughly investigated ATRP system is based on copper. It is catalyzed by copper salts complexed with bipyridines and other polydentate amines:

$$(5)$$

In agreement with some model reactions and estimates based on evolution of polydispersities with conversion, the deactivation process is very fast, nearly diffusion controlled, $k_d \approx 10^{7\pm1}$ M^{-1} s^{-1}. The equilibrium concentration of radicals is mainainted due to relatively fast activation process, $k_a \approx 10^{0\pm1}$ M^{-1} s^{-1}. The exact structure of Cu(I) and Cu(II) species in non-polar solvents is not precisely known, but X-ray analysis suggests that Cu(I) is in the form of a distorted tetrahedron *(33)* with the weakly coordinating anion and Cu(II) is a trigonal bipyramid *(34)* with one halogen atom in the inner sphere, and the other in the anionic form. Due to the low concentration of radicals, which is determined by the concentrations of both Cu(I) and Cu(II) species and dormant chains ([P°]= (k_a/k_d) [PX] [Cu(I)] / [Cu(II)]), the proportion of terminated chains is usually low (<5%).

ATRP has been successfully used for the controlled polymerization of various substituted styrenes,*(35)* acrylates,*(36)* acrylonitrile,*(37)* methacrylates,*(38)* dienes and some other monomers.*(39)* Molecular weight of the resulting polymers can be controlled usually in the range of M_n=1,000 to 100,000, although well defined dimers *(40)* as well as higher molecular weight polymers have also been prepared.*(38)* Polydispersities are usually well below M_w/M_n<1.5, which is the lower limit for conventional radical polymerizations, and polymers with polydispersities as low as M_w/M_n<1.05 have already been prepared.*(41)*

The termini of the macromolecule are defined by the initiator used. The tail end-group is generated by the alkylating part of the initiator which enables the introduction of a variety of functional groups such as allyl, hydroxy, epoxy, cyano, amino, vinyl and others.*(42)* The halogen atom from the initiator (or catalyst) occupies the head end-group of the macromolecule and can be easily displaced by azide and other nucleophiles.*(43)* Thus, ATRP is a very convenient method for the preparation of both symmetric and asymmetric telechelic macromolecules. Because radical polymerizations are tolerant to many functional groups, side functionalities have been incorporated by polymerization of hydroxyethyl (meth)acrylates, vinyl acrylate, trimethylsilyl (meth)acrylate and other functional monomers.

ATRP is a radical process and therefore it is possible to prepare various copolymers, much more easily than using ionic copolymerizations. The continuous growth of the chains with conversion is usually accompanied by the variation of the monomer feed composition in the simultaneous copolymerization. The drift in the monomer feed is reflected by the continuous changes in the composition of all chains resulting in the formation of gradient copolymers.*(44,45)* On the other hand, several segmented copolymers such as block and graft copolymers were prepared by sequential monomer additions.*(46,47)*

Finally, ATRP allows for controlled (co)polymer topology leading not only to linear chains but also to stars, combs and hyperbranched structures.*(47-49)* A brief summary of new (co)polymers prepared by ATRP is shown in Scheme 1 below.

Scheme 1
Polymers with New Topologies, Compositions and Functionalities Prepared by ATRP

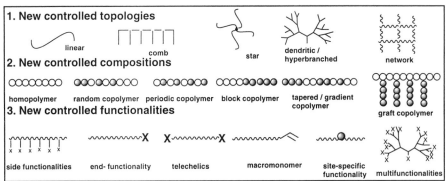

Special Features of Bulk Controlled Radical Polymerizations

There are several special features of bulk controlled radical polymerization. Probably the most important is the absence of the Trommsdorf (gel) effect. The others include chain length dependent termination, contribution of side reactions at high conversion and self-initiation, especially for styrene bulk polymerization. They will be discussed in this order.

Trommsdorf effect in bulk ATRP

In conventional radical polymerization first order kinetics in monomer is observed at low monomer conversion indicating a constant concentration of growing radicals. This stationary concentration is achieved by balancing slow initiation with termination by coupling/disproportionation between either two growing chains or a long growing chain and a newly formed radical (or short chain).(3) As polymerization progresses, monomer is consumed, and the viscosity increases due to progressively higher polymer concentrations. In some cases when the polymerization temperature is below the glass transition temperature of the formed polymer, vitrification can be observed resulting in incomplete monomer conversion due to a glass formation. When viscosity increases propagation rate constants decrease but termination rate constants decrease much more. This may result in an increase in the concentration of radicals, acceleration of the polymerization and the Trommsdorf or gel effect which may lead to an uncontrolled (explosive !) polymerization when heat transfer becomes inefficient. The main reason for the gel effect is that the rate of termination deceases much more than the rate of initiation. This difference originates in the increase of viscosity, but is progressively enhanced with temperature (energy of activation of the initiator decomposition is much higher than that of termination).

In controlled radical polymerizations the stationary concentration of radicals is established not due to equal rates of termination and initiation but due to a balance between rates of activation and deactivation. In ATRP, both bimolecular reactions include the polymer chain end and low molar mass species (activator Cu(I) and deactivator Cu(II), respectively). Both reactions should be affected by viscosity in the same way and identical to propagation, which also involves low molar mass monomer. It may happen that at the very end of the reaction, the diffusion control limit is significantly reduced and deactivation will become slower but this effect should be much smaller than for termination between two growing radicals. It has been reported before that the rate constant of termination/deactivation is very much chain length dependent.(50) At high conversion, the value of the termination rate constant between two macroradicals may be as low as $k_t \approx 10^2$ $mol^{-1} \cdot L \cdot s^{-1}$ though small radical and/or deactivator may react with a macroradical much faster, with $k_d \approx 10^7$ $mol^{-1} \cdot L \cdot s^{-1}$. Thus, the main reason for the absence of the Trommsdorf effect in both ATRP and nitroxide mediated bulk polymerization *(51)* is that rates of both propagation and deactivation are affected in a similar way by changes of viscosity.

However, a spontaneous increase of bulk polymerization temperature can be noted if polymerization is fast and the heat transfer at high conversion is not sufficient. The observed exotherm is usually related to the efficiency of heat transfer which will include volume of the reaction mixture, viscosity of the system (molecular weight, conversion, temperature), size and shape of the reactor, efficiency of mixing, etc.

Chain length dependent termination

Rate constants of termination in radical polymerization should be rather named rate coefficients because they vary with both conversion and chain length. There are several models for chain length dependent termination but the exact dependence is not yet experimentally established because in conventional radical systems chains are continuously generated and it is difficult to look at individual chains. Controlled bulk radical polymerization is very different because all chains start growing at the same time and the chain length increases linearly with conversion. This leads to several important features. With the progress of the reaction, viscosity increases due to progressively higher molecular weight, higher polymer concentration and consumption of solvent (monomer for bulk polymerization). Under such conditions termination coefficients decrease and the ratio of k_p/k_t increases, resulting in better control by reducing the contribution of chain breaking reactions. At the very beginning of polymerization, two initiating radicals may terminate with a very large rate constant, $k_t > 10^9$ mol^{-1}·L·s^{-1} ,(52) but this value quickly decreases to $k_t \approx 10^8$ mol^{-1}·L·s^{-1} and subsequently $k_t < 10^7$ mol^{-1}·L·s^{-1} . At the very beginning, small radicals terminate rapidly, produce an excess of deactivator (Cu(II) for ATRP or TEMPO for nitroxide mediated systems) and lead to a persistent radical effect.(53) At the same time some chains loose activity and functionality irreversibly. This effect can be partially avoided by adding an excess of deactivator at the beginning of the reaction. Additional improvement may be achieved by using oligomeric/polymeric macroinitiators. It is possible that careful analysis of the distribution of dead chains as a function of chain length will enable a better understanding of the effect of chain length on termination coefficients.

Side Reactions

Both ATRP and TEMPO-mediated styrene polymerization are well controlled up to $M_n \approx 20,000$. Polymers with good control of end groups and with polydispersities as low as $M_w/M_n < 1.05$ can be prepared. When the synthesis of higher molecular weight polymer is attempted, polymers with molecular weights lower than predetermined by $\Delta[M]/[I]_o$ and with progressively higher polydispersities are formed.(54,55) There are several reasons for these deviations from ideal behavior. They include both self-initiation (discussed in the next section) and other chain breaking reactions. The chain breaking reactions include formation of hydroxylamine and unsaturated end groups *(56)* and similar elimination of HX for ATRP.*(54)* It seems that by reducing the reaction temperature, it is possible to extend the range of molecular weights for ATRP of styrene. At 110 °C, the upper limit for the initiating system consisting of alkyl bromide and CuBr/bipy is in the range of $M_n \approx 80,000$, whereas at 100°C, polystyrenes with $M_n > 150,000$ were prepared. Additional improvement may be achieved by selecting catalysts which do not participate in outer sphere electron transfer reactions. For example, the use of $Cu(CH_3CN)_4$ leads predominantly to carbocationic polymerization of styrene, *(57)* whereas $Cu(bipy)_2$ results in radical polymerization.*(58)* Both oxidation of growing radicals and heterolytic cleavage of the R-X bond can happen in the former case.

It should be stressed that side reactions are kinetically zero order with respect to monomer and their contribution becomes more important at higher conversion and at higher dilution. Therefore, bulk polymerization often provides better controlled polymers than solution polymerization.*(58)*

Contribution of Self-Initiation in Styrene Bulk Polymerization

In bulk polymerization of styrene and substituted styrenes, thermal self-initiated polymerization of styrene may occur simultaneously with the controlled process. The rates of thermal self-initiation were measured at various temperatures.*(59-61)* The overall rate of self-initiated polymerization should not be confused with the rate of radical generation. For example, approximately 1 M/hour (about 15%/hour) of styrene is consumed at 130 °C in bulk thermal polymerization, but less than 10^{-3} M/hour of radicals are generated. Thus, in 10 hours less than 10^{-2} M radicals (or new chains) are generated. This number may even be lower because the rate of self-initiation decreases with conversion since it is second or third order with respect to monomer concentration.*(60,62)*

It has been reported that in some TEMPO and other nitroxide-mediated styrene polymerizations, the overall polymerization rate is very similar to the rate of the thermal self-initiated process.*(16,63,64)* This behavior was attributed to the relatively small contribution of radicals generated by activation (dissociation) of dormant alkoxyamines to the overall propagation rate.*(16)* Nevertheless, molecular weights in nitroxide mediated systems are much smaller than in the thermal process and correlate relatively well with the amount of TEMPO or alkoxyamine present in the system ($DP_n \approx \Delta[M]/[I]_o$). This indicates that thermally generated chains are continuously trapped by the small excess of TEMPO present in the system and the <u>total</u> concentration of chains formed by thermal self-initiation is small in comparison with nitroxide-capped chains. This system can be considered self-regulating, because without self-initiation the overall polymerization rate would be much smaller, as indeed observed for polymerizations of acrylates. The overall polymerization rate is defined by the total concentration of radicals, those resulting from self-initiation and those from the reversible activation process. Since the exchange between them is fast, they all grow simultaneously, yielding well-defined polymers. Because the equilibrium constant in TEMPO mediated styrene polymerization is low, e.g. $K=10^{-11}$ M, the contribution of radicals generated by dissociation of alkoxyamines is estimated to be less than 30% of the total concentrations of radicals when $[P\text{-}X]_o=10^{-2}$M, but this value is expected to reach 50% when $[P\text{-}X]_o=10^{-1}$ M. Thus, potentially, at still higher concentration of alkoxyamines the rate of polymerization could start increasing, as shown in simulated kinetic plots in Figure 1.

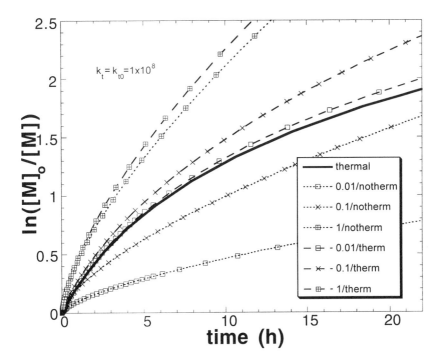

Fig. 1. Simulated kinetics of bulk styrene polymerization mediated by TEMPO-based alkoxyamine at 130 °C. Simulations use value $K=2.5 \ 10^{-11}$M. Kinetic parameters for the thermal self-initiation were taken from Ref. *(16)*. The concentration of alkoxyamines was varied: 1, 0.1, 0.01 M.

Figure 1 (and 2) shows the rate of thermal styrene polymerization at 130 °C (solid line) together with the rates of TEMPO-mediated bulk styrene polymerizations at various concentrations of alkoxyamines. An anticipated rate was calculated for hypothetical systems without thermal self initiation. The overal rates at $[P-X]_o=10^{-1}$ and 10^{-2} M are very close to that in the thermal self-initiated system and the hypothetical rates without thermal self initiation were correspondingly lower.

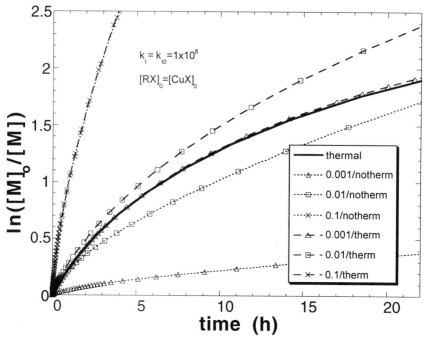

Fig. 2. Kinetics of bulk ATRP of styrene at 130 °C. Simulations are based on K=3 x 10^{-8}. $[CuCl/2dNbipy]_o = [RCl]_o = 0.1, 0.01$ and 0.001M.

However, at $[P-X]_o=1$ M the overall polymerization rate with alkoxyamines was expected to be larger than for a thermally self-initiated process. Nonetheless, experiments do not confirm this prediction and indicate an even slower rate at higher alkoxyamine concentration. A plausible explanation may be due to the variation of the termination coefficients with the chain length. The initiating radicals (e.g. 1-phenylethyl) terminate with rate constants of k=5 x 10^9 M^{-1} s^{-1}, which is 50 times faster than for similar reactions involving polymeric radicals. This rapid termination may lead to the formation of excess nitroxide at the early stages and to the reduction of the overall polymerization rate. Another approach to enhance the proportion of radicals formed from alkoxyamines is to increase the value of the equilibrium constant by either increasing temperature or using alkoxyamines which dissociate faster. Unfortunately, increasing temperature does not improve the system because both dissociation and thermal self-initiation are accelerated. However, the second approach based on less stable alkoxyamines was quite successfully applied to polymerization of acrylates.*(65)*

Similar behavior can be observed in ATRP. Though, when the concentration of alkyl halides and CuX/2bipy are high enough (both around 0.1 M), the polymerization rate is >10 times faster than the rate of thermal self-initiated polymerization. This is due to a much higher equilibrium constant for ATRP (K=3 x 10^{-8} at 130 °C for RCl/CuCl and an additional ten times higher for RBr/CuBr system).*(58)* However, when the concentration of either catalyst or initiator is significantly reduced (ATRP is first order with respect to initiator and catalyst), the overall polymerization rate becomes comparable to that of a thermal process. A further decrease in the concentration of either reagent does not additionally reduce the polymerization rate which, under steady-state conditions, can not be smaller than that of a thermal self-initiation process. This minimum rate defined by thermal self-initiation is not observed for (meth)acrylates which do not self-initiate.

Self-initiation is important for the polymerization of styrene. It controls the polymerization rates for most TEMPO-mediated systems as well as some ATRP systems when low concentrations of either catalyst or initiator are used. Nevertheless, the exchange reactions enable satisfactory control of molecular weights in both systems.

Conclusions

Bulk controlled radical polymerizations were successfully used for the synthesis of polymers with controlled molecular weights, end functionalities and low polydispersities, as well as controlled chain architecture and composition. Atom transfer radical polymerization is a robust controlled radical system which has been successfully applied to bulk polymerization of substituted styrenes and functional (meth)acrylates. Bulk controlled radical polymerizations are characterized by the absence of the Trommsdorf effect, chain length dependent termination, and the dependence of side reactions such as decomposition of growing chains and self-initiation on temperature and monomer concentration.

Acknowledgments. Support from the ATRP Consortium is gratefully acknowledged.

References

(1) Szwarc, M. *Nature* 1956, *176*, 1168.
(2) Greszta, D.; Mardare, D.; Matyjaszewski, K. *Macromolecules* 1994, *27*, 638.

(3) Moad, G.; Solomon, D. H. *The Chemistry of Free Radical Polymerization*; Pergamon: Oxford, 1995.

(4) Matyjaszewski, K.; Gaynor, S.; Wang, J. S. *Macromolecules* 1995, *28*, 2093.

(5) Gaynor, S.; Wang, J. S.; Matyjaszewski, K. *Macromolecules* 1995, *28*, 8051.

(6) Yutani, Y.; Tatemoto, M. In *European Patent #0 489 370 A1*; Europ. Pat., 1991; pp 1.

(7) Moad, C. L.; Moad, G.; Rizzardo, E.; Tang, S. H. *Macromolecules* 1996, *29*, 7717.

(8) Greszta, D.; D. Mardare; Matyjaszewski, K. *Polym. Prepr. (Am. Chem. Soc., Div. Polym. Chem.)* 1994, *35(1)*, 466.

(9) Harwood, H. J.; Christov, L.; Guo, M.; Holland, T. V.; Huckstep, A. Y.; Jones, D. H.; Medsker, R. E.; Rinaldi, P. L.; Soito, T.; Tung, D. S. *Macromol. Symp.* 1996, *111*, 25.

(10) Mardare, D.; Matyjaszewski, K. *Macromolecules* 1994, *27*, 645.

(11) Minoura, Y.; Utsumi, K.; Lee, M. *J. Chem. Soc. Faraday Trans. 1* 1979, *75*, 1821.

(12) Rizzardo, E. *Chem. Aust.* 1987, *54*, 32.

(13) Georges, M. K.; Veregin, R. P. N.; Kazmaier, P. M.; Hamer, G. K. *Macromolecules* 1993, *26*, 2987.

(14) Gaynor, S.; Greszta, D.; Mardare, D.; Teodorescu, M.; Matyjaszewski, K. *J. Macromol. Sci., Pure Appl. Chem.* 1994, *A31*, 1561.

(15) Hawker, C. J. *J. Am. Chem. Soc.* 1994, *116*, 11185.

(16) Greszta, D.; Matyjaszewski, K. *Macromolecules* 1996, *29*, 7661.

(17) Otsu, T.; Yoshida, M. *Makromol. Chem. Rapid Commun.* 1982, *3*, 127.

(18) Borsig, E.; Lazar, M.; Capla, M.; Florian, S. *Angew. Makromol. Chem.* 1969, *9*, 89.

(19) Braun, D. *Macromol. Symp.* 1996, *111*, 63.

(20) Lee, M.; Utsumi, K.; Minoura, Y. *J. Chem. Soc., Farad. Trans. 1* 1979, *75*, 1821.

(21) Wayland, B. B.; Poszmik, G.; Mukerjee, S. L.; Fryd, M. *J. Am. Chem. Soc.* 1994, *116*, 7943.

(22) Harwood, H. J.; Arvanitopoulos, L. D.; Greuel, M. P. *Polym. Prepr. (Am. Chem. Soc., Div. Polym. Chem.)* 1994, *35(2)*, 549.

(23) Davis, T. P.; Haddleton, D. M.; Richards, S. N. *J. Macromol. Sci., Rev. Macromol. Chem. Phys.* 1994, *C34*, 243.

(24) Turner, R. S.; Blevins, R. W. *Macromolecules* 1990, *23*, 1856.

(25) Wang, J. S.; Matyjaszewski, K. *J. Am. Chem. Soc.* 1995, *117*, 5614.

(26) Kato, M.; Kamigaito, M.; Sawamoto, M.; Higashimura, T. *Macromolecules* 1995, *28*, 1721.

(6) Yutani, Y.; Tatemoto, M. In *European Patent #0 489 370 A1*; Europ. Pat., 1991; pp 1.

(7) Moad, C. L.; Moad, G.; Rizzardo, E.; Tang, S. H. *Macromolecules* 1996, *29*, 7717.

(8) Greszta, D.; D. Mardare; Matyjaszewski, K. *Polym. Prepr. (Am. Chem. Soc., Div. Polym. Chem.)* 1994, *35(1)*, 466.

(9) Harwood, H. J.; Christov, L.; Guo, M.; Holland, T. V.; Huckstep, A. Y.; Jones, D. H.; Medsker, R. E.; Rinaldi, P. L.; Soito, T.; Tung, D. S. *Macromol. Symp.* 1996, *111*, 25.

(10) Mardare, D.; Matyjaszewski, K. *Macromolecules* 1994, *27*, 645.

(11) Minoura, Y.; Utsumi, K.; Lee, M. *J. Chem. Soc. Faraday Trans. 1* 1979, *75*, 1821.

(12) Rizzardo, E. *Chem. Aust.* 1987, *54*, 32.

(13) Georges, M. K.; Veregin, R. P. N.; Kazmaier, P. M.; Hamer, G. K. *Macromolecules* 1993, *26*, 2987.

(14) Gaynor, S.; Greszta, D.; Mardare, D.; Teodorescu, M.; Matyjaszewski, K. *J. Macromol. Sci., Pure Appl. Chem.* 1994, *A31*, 1561.

(15) Hawker, C. J. *J. Am. Chem. Soc.* 1994, *116*, 11185.

(16) Greszta, D.; Matyjaszewski, K. *Macromolecules* 1996, *29*, 7661.

(17) Otsu, T.; Yoshida, M. *Makromol. Chem. Rapid Commun.* 1982, *3*, 127.

(18) Borsig, E.; Lazar, M.; Capla, M.; Florian, S. *Angew. Makromol. Chem.* 1969, *9*, 89.

(19) Braun, D. *Macromol. Symp.* 1996, *111*, 63.

(20) Lee, M.; Utsumi, K.; Minoura, Y. *J. Chem. Soc., Farad. Trans. 1* 1979, *75*, 1821.

(21) Wayland, B. B.; Poszmik, G.; Mukerjee, S. L.; Fryd, M. *J. Am. Chem. Soc.* 1994, *116*, 7943.

(22) Harwood, H. J.; Arvanitopoulos, L. D.; Greuel, M. P. *Polym. Prepr. (Am. Chem. Soc., Div. Polym. Chem.)* 1994, *35(2)*, 549.

(23) Davis, T. P.; Haddleton, D. M.; Richards, S. N. *J. Macromol. Sci., Rev. Macromol. Chem. Phys.* 1994, *C34*, 243.

(24) Turner, R. S.; Blevins, R. W. *Macromolecules* 1990, *23*, 1856.

(25) Wang, J. S.; Matyjaszewski, K. *J. Am. Chem. Soc.* 1995, *117*, 5614.

(26) Kato, M.; Kamigaito, M.; Sawamoto, M.; Higashimura, T. *Macromolecules* 1995, *28*, 1721.

(27) Matyjaszewski, K.; Wei, M.; Xia, J.; McDermott, N. E. *Macromolecules* 1997, *30*, 8161.

(28) Ando, T.; Kamigaito, M.; Sawamoto, M. *Macromolecules* 1997, *30*, 4507.

(29) Granel, C.; Dubois, P.; Jerome, R.; Teyssie, P. *Macromolecules* 1996, *29*, 8576.

110

(27) Matyjaszewski, K.; Wei, M.; Xia, J.; McDermott, N. E. *Macromolecules* 1997, *30*, 8161.

(28) Ando, T.; Kamigaito, M.; Sawamoto, M. *Macromolecules* 1997, *30*, 4507.

(29) Granel, C.; Dubois, P.; Jerome, R.; Teyssie, P. *Macromolecules* 1996, *29*, 8576.

(30) Percec, V.; Barboiu, B.; Neumann, A.; Ronda, J. C.; Zhao, M. *Macromolecules* 1996, *29*, 3665.

(31) Uegaki, H.; Kotani, Y.; Kamigaito, M.; Sawamoto, M. *Macromolecules* 1997, *30*, 2249.

(32) Curran, D. P. *Synthesis* 1988, 489.

(33) Munakata, M.; Kitagawa, S.; Asahara, A.; Masuda, H. *Bull. Chem. Soc. Jpn.* 1987, *60*, 1927.

(34) Hathaway, B. J. In *Comprehensive Coordination Chemistry*; G. Wilkinson, R. D. Gillard and J. A. McClaverty, Eds.; Pergmanon: Oxford, 1987; Vol. 5; pp 533.

(35) Qiu, J.; Matyjaszewski, K. *Macromolecules* 1997, *30*, 5643.

(36) Coca, S.; Matyjaszewski, K. *Macromolecules* 1997, *30*, 2808.

(37) Matyjaszewski, K.; Jo, S.; Paik, H.-J.; Gaynor, S. *Macromolecules* 1997, *30*, 6398.

(38) Grimaud, T.; Matyjaszewski, K. *Macromolecules* 1997, *30*, 2216.

(39) Matyjaszewski, K.; Wang, J. S. *WO 96/30421* 1996,

(40) Bellus, D. *Pure Appl. Chem.* 1985, *57*, 1827.

(41) Patten, T. E.; Xia, J.; Abernathy, T.; Matyjaszewski, K. *Science* 1996, *272*, 866.

(42) Matyjaszewski, K.; Coca, S.; Nakagawa, Y.; Xia, J. *Polym. Mat. Sci. Eng.* 1997, *76*, 147.

(43) Nakagawa, Y.; Gaynor, S. G.; Matyjaszewski, K. *Polym. Prepr. (Am. Chem. Soc., Div. Polym. Chem.)* 1996, *37(1)*, 577.

(44) Pakula, T.; Matyjaszewski, K. *Macromol. Theory & Simulat.* 1996, *5*, 987.

(45) Greszta, D.; K.Matyjaszewski *Polym. Prepr. (Am. Chem. Soc., Div. Polym. Chem.)* 1996, *37(1)*, 569.

(46) Jo, S. M.; Gaynor, S. G.; Matyjaszewski, K. *Polym. Prepr. (Am. Chem. Soc., Div. Polym. Chem.)* 1996, *37(2)*, 272.

(47) Wang, J. S.; Greszta, D.; Matyjaszewski, K. *Polym. Mater. Sci. Eng.* 1995, *73*, 416.

(48) Gaynor, S. G.; Edelman, S.; Matyjaszewski, K. *Macromolecules* 1996, *29*, 1079.

(49) Matyjaszewski, K.; Gaynor, S.; Kulfan, A.; Podwika, M. *Macromolecules* 1997, *30*, 5192.

(50) Scheren, P. A. G. M.; Russell, G. T.; Sangster, D. F.; Gilbert, R. G.; German, A. L. *Macromolecules* 1995, *28*, 3637.

(30) Percec, V.; Barboiu, B.; Neumann, A.; Ronda, J. C.; Zhao, M. *Macromolecules* 1996, *29*, 3665.

(31) Uegaki, H.; Kotani, Y.; Kamigaito, M.; Sawamoto, M. *Macromolecules* 1997, *30*, 2249.

(32) Curran, D. P. *Synthesis* 1988, 489.

(33) Munakata, M.; Kitagawa, S.; Asahara, A.; Masuda, H. *Bull. Chem. Soc. Jpn.* 1987, *60*, 1927.

(34) Hathaway, B. J. In *Comprehensive Coordination Chemistry*; G. Wilkinson, R. D. Gillard and J. A. McClaverty, Eds.; Pergmanon: Oxford, 1987; Vol. 5; pp 533.

(35) Qiu, J.; Matyjaszewski, K. *Macromolecules* 1997, *30*, 5643.

(36) Coca, S.; Matyjaszewski, K. *Macromolecules* 1997, *30*, 2808.

(37) Matyjaszewski, K.; Jo, S.; Paik, H.-J.; Gaynor, S. *Macromolecules* 1997, *30*, 6398.

(38) Grimaud, T.; Matyjaszewski, K. *Macromolecules* 1997, *30*, 2216.

(39) Matyjaszewski, K.; Wang, J. S. *WO 96/30421* 1996,

(40) Bellus, D. *Pure Appl. Chem.* 1985, *57*, 1827.

(41) Patten, T. E.; Xia, J.; Abernathy, T.; Matyjaszewski, K. *Science* 1996, *272*, 866.

(42) Matyjaszewski, K.; Coca, S.; Nakagawa, Y.; Xia, J. *Polym. Mat. Sci. Eng.* 1997, *76*, 147.

(43) Nakagawa, Y.; Gaynor, S. G.; Matyjaszewski, K. *Polym. Prepr. (Am. Chem. Soc., Div. Polym. Chem.)* 1996, *37(1)*, 577.

(44) Pakula, T.; Matyjaszewski, K. *Macromol. Theory & Simulat.* 1996, *5*, 987.

(45) Greszta, D.; K.Matyjaszewski *Polym. Prepr. (Am. Chem. Soc., Div. Polym. Chem.)* 1996, *37(1)*, 569.

(46) Jo, S. M.; Gaynor, S. G.; Matyjaszewski, K. *Polym. Prepr. (Am. Chem. Soc., Div. Polym. Chem.)* 1996, *37(2)*, 272.

(47) Wang, J. S.; Greszta, D.; Matyjaszewski, K. *Polym. Mater. Sci. Eng.* 1995, *73*, 416.

(48) Gaynor, S. G.; Edelman, S.; Matyjaszewski, K. *Macromolecules* 1996, *29*, 1079.

(49) Matyjaszewski, K.; Gaynor, S.; Kulfan, A.; Podwika, M. *Macromolecules* 1997, *30*, 5192.

(50) Scheren, P. A. G. M.; Russell, G. T.; Sangster, D. F.; Gilbert, R. G.; German, A. L. *Macromolecules* 1995, *28*, 3637.

(51) Saban, M. D.; Georges, M. K.; Veregin, R. P. N.; Hamer, G. K.; Kazmaier, P. M. *Macromolecules* 1995, *28*, 7032.

(52) Fischer, H.; Paul, H. *Acc. Chem. Res.* 1987, *20*, 200.

(53) Fischer, H. *J. Am. Chem. Soc.* 1986, *108*, 3925.

(51) Saban, M. D.; Georges, M. K.; Veregin, R. P. N.; Hamer, G. K.; Kazmaier, P. M. *Macromolecules* 1995, *28*, 7032.

(52) Fischer, H.; Paul, H. *Acc. Chem. Res.* 1987, *20*, 200.

(53) Fischer, H. *J. Am. Chem. Soc.* 1986, *108*, 3925.

(54) Matyjaszewski, K.; Davis, K.; Patten, T., Wei, M. *Tetrahedron*, 1997, *53*, 15321.

(55) Zhu, Y.; Howell, B. A.; Priddy, D. B. *Am. Chem. Soc. Polym. Preprints* 1997, *38(1)*, 97.

(56) Li, I.; Howell, B. A.; Matyjaszewski, K.; Shigemoto, T.; Smith, P. B.; Priddy, D. B. *Macromolecules* 1995, *28*, 6692.

(57) Haddleton, D. M.; Shooter, A. J.; Hannon, M. J.; Barker, J. A. *Polym. Prepr. (Am. Chem. Soc., Div. Polym. Chem.)* 1997, *38(1)*, 679.

(58) Matyjaszewski, K.; Patten, T.; Xia, J. *J. Am. Chem. Soc.* 1997, *119*, 674.

(59) Russell, K. E.; Tobolsky, A. V. *J. Am. Chem. Soc.* 1953, *75*, 5052.

(60) Hui, A. W.; Hamielec, A. E. *J. Appl. Polym. Sci.* 1972, *16*, 749.

(61) Barr, N. J.; Bengough, W. I.; Beveridge, G.; Park, G. B. *Europ. Polym. J.* 1978, *14*, 245.

(62) Fukuda, T.; Terauchi, T. *Chem. Lett.* 1996, 293.

(63) Catala, J. M.; Bubel, F.; Hammouch, S. O. *Macromolecules* 1995, *28*, 8441.

(64) Fukuda, T.; Terauchi, T.; Goto, A.; Ohno, K.; Tsujii, Y.; Miyamoto, T.; Kobatake, S.; Yamada, B. *Macromolecules* 1996, *29*, 6393.

(65) Benoit, D.; Grimaldi, S.; Finet, J. P.; Tordo, P.; Fontanille, M.; Gnanou, Y. *Polym. Prepr. (Am. Chem. Soc., Div. Polym. Chem.)* 1997, *38(1)*, 729.

Chapter 7

Bulk Free Radical Copolymerization of Allylic Alcohol with Acrylate and Styrene Comonomers

Shao-Hua Guo

ARCO Chemical Company, 3801 West Chester Pike, Newtown Square, PA 19073

The bulk free radical copolymerization of allylic alcohol with acrylate leads to new hydroxyl functional acrylic resins. The copolymerization does not require a process solvent and produces oligomers without need of chain transfer agent. The comonomer feed composition determines not only the copolymer composition but also the molecular weight of the copolymer and the polymerization rate. An excess of allylic alcohol is applied to achieve the high hydroxyl content in the copolymer, and the unreacted allyl monomer is removed and recycled. The copolymers have considerably more uniform hydroxyl group distribution along a polymer chain and among the polymer chains due to the low monomeric reactivity ratio of the allyl monomer. The hydroxyl acrylic resins have potential in automotive coatings and many other applications.

Allyl alcohol, a well-known monomer, is commercially available from the isomerization of propylene oxide. Free radical copolymerization of allyl alcohol with other vinyl monomers is a potential route to hydroxyl functional polymers. Such polymers are valuable intermediates for coatings, elastomers, and other thermoset polymers because they can be readily cured with multiple-functional isocyanates, anhydrides, melamine, and many other crosslinking agents.

Allyl alcohol readily reacts with an alkylene oxide [1] to form an alkoxylated allyl alcohol such as propoxylated (R: CH_3) or ethoxylated (R: H) allyl alcohol (Equation 1). Alkoxylated allyl alcohol has lower acute toxicity and lower vapor

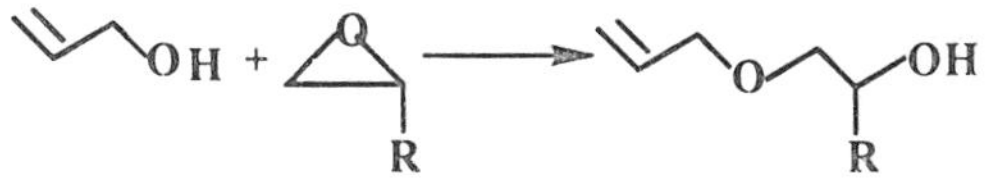

$$\text{(1)}$$

pressure than allyl alcohol, and, therefore, is easier to handle in the preparation of polymers [2]. The physical properties of allyl alcohol and allyl propoxylates are listed in Table I.

Table I. Physical Properties of Allyl Alcohol and Allyl Propoxylate Monomers

Monomer	Allyl Alcohol	Allyl Monopropoxylate	Allyl Propoxylate AAP 1.6
Structure			
Molecular Weight	58	116	150
Boiling Point, oC	97	145	145-245
Flash Point, oC	21	Not Available	63
LD 50, Oral mg/Kg	65	Not Available	1,100
Homopolymer Glass Transition Temperature	4 oC	-15 oC	-33oC

The allylic alcohols, including allyl alcohol and allyl alkoxylate, readily undergo free radical copolymerization with vinyl comonomers such as styrene, acrylate, and vinyl ether or ester [3-5]. For example, allyl alcohol and styrene free radically copolymerize to form the SAA copolymer (see equation 2) which is a unique class of resinous polyol for the coating industry [6-8]. Three SAA copolymers are available. They are different in the composition and the physical properties such as the hydroxyl number, the molecular weight and the glass transition temperature (Table II).

Table II.   Compositions and Physical Properties of SAA Copolymers

Product Name	SAA-103	SAA-100	SAA-101
Allyl Alcohol/Styrene Mole Ratio	20:80	30:70	40:60
Mw (GPC)	8,400	3,000	2,500
Mn (GPC)	3,200	1,500	1,200
Hydroxy Number mgKOH/g	125	210	255
Glass Transition Temperature, T_g, oC	78	62	57

$$\underset{\substack{\text{CH}_2 \\ | \\ \text{O H}}}{\text{CH}_2=\text{CH}} + \underset{\text{C}_6\text{H}_5}{\text{CH}_2=\text{CH}} \longrightarrow \underset{\substack{\text{CH}_2 \\ | \\ \text{O H}}}{\text{mCH}_2-\text{CH}-\text{CH}_2-\text{CH}m} \quad (2)$$

RESULTS AND DISCUSSION

Bulk Free Radical Copolymerization Process

The copolymerization of the allylic alcohol with acrylate and styrene was conducted in a semi-batch non-solvent process. All of the allylic alcohol and part of the acrylate and styrene comonomers were initially fed into the reactor, and the remaining acrylate and styrene comonomers were added during the polymerization into the reactor at such a rate as to maintain an essentially constant ratio of the allylic alcohol to the acrylate and styrene comonomers. High polymerization temperature (120°C to 160°C) was applied to achieve commercially acceptable monomer conversion and polymerization rate. The polymerization took 5 to 8 hours to reach 70 to 95% of the total monomer conversion. The unreacted monomers containing mostly the allylic alcohol (larger than 95%) and a small amount of acrylate and styrene comonomers and the initiator were removed after the polymerization. The copolymerization was carried out under the vapor pressure of the allylic alcohol. For example, the vapor pressure of allyl alcohol at 135°C is about 2.41 x 10^5 Pa. Di-t-butyl peroxide was used as the free radical initiator due to its high decomposition temperature. The initiator was gradually fed into the reactor mixture to maintain a steady ratio of initiator to the monomer and to achieve an increased monomer conversion [9].

A typical laboratory procedure for conducting the copolymerization of the allylic alcohol with the acrylate and styrene comonomers is described in the Experimental section. In order to design the semi-batch copolymerization process, we have investigated the bulk copolymerizations of the allylic alcohol, including allyl alcohol and allyl propoxylate, with methyl methacrylate and styrene, respectively. We measured the monomeric reactivity ratios and the copolymerization rates.

Monomeric Reactivity Ratios and Copolymer Compositions. The initial comonomer feed composition and the addition rate of the acrylate and styrene comonomers are determined by the desired hydroxyl group content (or hydroxyl number) of the copolymer, and by the monomeric reactivity ratios of the allylic alcohol and the acrylate and styrene comonomers. Most studies on allyl monomer polymerization or copolymerization in the literature were conducted at low polymerization temperature [10], and, therefore, they are of limited value for the commercial process. We experimentally measured the monomeric reactivity ratios of allyl alcohol and allyl propoxylate in bulk copolymerization with methyl methacrylate and styrene, respectively, at 135°C. The method known in the literature was applied for measuring the monomeric reactivity ratios [11]. For example, by determining the copolymer compositions of allyl alcohol with methyl methacrylate for several different comonomer feed compositions (Table III) and plotting the left side of the copolymer composition equation (Equation 3) against the coefficient of r_1, we obtained slope r_1 (monomeric reactivity ratio of methyl methacrylate) and intercept r_2 (monomeric reactivity ratio of allyl alcohol). The monomeric reactivity ratios of three monomer pairs are listed in Table IV.

Table III. Experimental Determination of the Monomeric Reactivity Ratios of Allyl Alcohol and Methyl Methacrylate @ 135°C

$f_1 \times 100$	4.94	7.24	9.42	17.3
$F_1 \times 100$	59.2	66.6	73.6	81.9
$\dfrac{f_1(1-2F_1)}{(1-f_1)F_1}$	-1.62×10^{-2}	-3.89×10^{-2}	-6.67×10^{-2}	-16.3×10^{-2}
$\dfrac{f_1^2(F_1-1)}{(1-f_1)^2 F_1}$	-1.86×10^{-3}	-3.06×10^{-3}	-3.88×10^{-3}	-9.67×10^{-3}

$$\frac{f_1(1-2F_1)}{(1-f_1)F_1} = r_2 + \frac{f_1^2(F_1-1)}{(1-f_1)^2 F_1} r_1 \tag{3}$$

where f_1 is the molar fraction of methyl methacrylate in the total monomers, and F_1 the molar fraction of methyl methacrylate monomeric unit in the copolymer being formed at any instant.

$$f_1 = \frac{[MMA]}{[MMA]+[AA]} \qquad F_1 = \frac{d[MMA]}{d[MMA]+d[AA]} \tag{4}$$

As the monomeric reactivity ratio of allylic alcohol is dramatically lower than that of the acrylate or styrene, a large ratio of the allylic alcohol to the acrylate or styrene comonomer is always needed to produce a resin of high hydroxyl group content. The excess amount of the allylic alcohol is removed after the polymerization by the vacuum distillation. The amount of allylic alcohol recycled depends on the desired hydroxyl content of the copolymer product, and it can be predicted by the copolymer composition equation. It is usually in a range of 5-30% of the total reactor charge for preparing a copolymer containing 10-50 molar % of the allylic alcohol.

Table IV. Monomeric Reactivity Ratios of Allyl Alcohol and Allyl Propoxylate in Bulk Radical Copolymerization with Methyl Methacrylate and Styrene, respectively, at 135°C

M_1	M_2	r_1	r_2
Methyl Methacrylate	Allyl Alcohol	19.0	0.02
Methyl Methacrylate	Allyl Propoxylate	11.7	0.05
Styrene	Allyl Alcohol	69.0	0.05

Copolymerization Kinetics. The rate expression of free radical copolymerization is rather complex and, therefore, few rate constants of copolymerizations are measurable. The copolymerization of the allylic alcohol with acrylate or styrene provides a simpler system for evaluating the copolymerization kinetics since the allylic alcohol hardly undergoes homopolymerization (r_2 close to zero, see Table IV). For example, the rate expression of styrene and allyl alcohol copolymerization can be approximated by (internal technical report of ARCO Chemical Company)

$$R_p = \left([St] + \frac{2[AA]}{r_1} \right) K [I]^{1/2} \tag{5}$$

$$K = \left(\frac{fk_d}{k_{t11}} \right)^{1/2} k_{11} \tag{6}$$

where K is the apparent rate constant, f the initiator efficiency, k_d the initiator decomposition rate constant, k_{11} the rate constant for a propagating chain ending in styrene adding to styrene monomer (ie., homopolymerization rate constant), and k_{t11} the chain termination rate constant for two propagating chains ending in styrene. By measuring the copolymerization rate R_p for several different concentrations of styrene, allyl alcohol and initiator at a low degree of monomer conversion ($< 5\%$) (Table V) and plotting R_p against $([St] + 2[AA]/r_1)[I]^{1/2}$, we obtained a straight line which had slop 0.0328 and intercept 6.5×10^{-5} (theoretical value is zero). Therefore, the apparent rate constant of copolymerization of allyl alcohol and styrene at 135°C $K = 0.0328$ mole$^{-1/2}$. min$^{-1/2}$. l.

Similarly, the copolymerization kinetics of allyl alcohol and allyl propoxylate, respectively, with methyl methacrylate have been studied. According to the rate constants, we can calculate the consumption rates of the allylic alcohol and the acrylate and styrene comonomers, and determine the acrylate and styrene comonomer addition rate to keep the ratio of the allylic alcohol to the acrylate and styrene comonomers essentially constant. Equation 5 also predicts the polymerization heat which needs to be removed from the polymerization system.

Table V. Evaluation Of Copolymerization Rate Constant

Exp No	[St] mole/l	[AA] mole/l	[I] mole/l	$([St] + 2[AA]/r_1)[I]^{1/2}$	R_p
1	0.552	11.02	0.1410	0.328	0.012
2	0.563	11.27	0.0448	0.189	0.0082
3	1.049	10.50	0.0417	0.277	0.0083
4	0.817	10.87	0.0432	0.236	0.0056

We noticed that the apparent copolymerization rate constant K contains only the rate constants related to styrene homopolymerization. The typical rate expression of the styrene homopolymerization is given by

$$R_p = k_p \left(\frac{fk_d}{k_t} \right)^{1/2} [St]\ [I]^{1/2} = K^o [St]\ [I]^{1/2}$$

(7)

where K^o is the apparent rate constant for styrene homopolymerization, k_p the propagation rate constant, and k_t the chain termination rate constant. According to the literature data [12-13] of k_p, k_d, k_t, and f, we calculated K^o value for the free radical homopolymerization at 135 ºC, and $K^o = k_p(fk_d/k_t)^{1/2} = 0.0398$ mol$^{-1/2}$ min$^{-1/2}$ l . The apparent rate constant of styrene homopolymerization K^o nicely matches K, the apparent rate constant of the copolymerization of styrene with allyl alcohol. From the Equation 5, we see that the contribution of allyl alcohol to the copolymerization rate is small although its concentration in the monomer mixture may be large. The polymerization heat, therefore, is mainly contributed by the polymerization of styrene. The initial reactor charge of the semi-batch process contains about 15-35% of the acrylate and styrene comonomers and 65-85% of the allylic alcohol. So the requirement for the heat removal in the copolymerization of the allylic alcohol with the acrylate and styrene is very close to the solution polymerization of acrylate and styrene in the same monomer concentration. The excess allylic alcohol which is required to achieve the desired hydroxyl content of the copolymer functions as a solvent in the polymerization process for controlling the polymerization rate and removing the polymerization heat.

Characteristics of the Copolymers of Allylic Alcohol with Acrylate and Styrene

The copolymers of the allylic alcohol with the acrylate and styrene comonomers are of many unique characteristics. They include the dependence of molecular weight on the copolymer composition, the consistent hydroxyl functionality (the number of hydroxyl functional group per polymer chain) among the copolymer chains, and the low concentration of the blocks of the hydroxyl functional monomeric units along a polymer chain. These copolymers are expected to offer more uniform crosslinking networks, and, therefore, to improve the performance of the coatings and other thermoset polymer products.

Molecular Weight and Hydroxyl Functionality. The dependence of molecular weight on the copolymer composition was studied under low-conversion polymerization (less than 5% total monomer conversion) with various comonomer feed compositions. The number average molecular weights of the copolymers have been found to be in a range of 1,000 to 10,000 in a broad range of copolymer compositions. The higher the concentration of the allylic monomer in the copolymer, the lower the molecular weight (Table VI-VIII).

The generally accepted mechanism for the formation of low molecular weight in the polymerization of allyl monomers is the *degradative chain transfer* [14], that is, the propagating radical transfers to the allyl monomer to form the allylic radical which is too weak to grow. However, this was not agreed by the ESR studies which indicated that the allylic radicals in the polymerizations of allyl or methallyl monomers were very low [15].

Very interestingly, we have found that the hydroxyl functionalities of the copolymers were dependent neither on the molecular weights of the copolymers nor on the copolymer compositions. They were determined by the relative reactivities of the comonomers. Allyl propoxylate is relatively more reactive than allyl alcohol, and its copolymers with methyl methacrylate have a higher hydroxyl functionality (the average functionality is about 12, Table VII) than the copolymers of allyl alcohol

Table VI. Dependencies of Molecular Weight and Functionality on Composition of Allyl Alcohol and Methyl Methacrylate Copolymers*

Allyl Alcohol/MMA in mole	18:82	26:74	33:67	41:59
Mw	13040	5970	4640	3350
Mn	5110	2690	2160	1690
Mw/Mn	2.55	2.22	2.15	1.98
Number Average Functionality	9.0	7.9	8.2	8.5

* Polymers prepared at 135°C

with methyl methacrylate (the average functionality is about 8, Table VI). On other

120

side, methyl methacrylate is more reactive than styrene, and the hydroxyl functionality of its copolymers with allyl alcohol is significantly higher than the copolymers of styrene with allyl alcohol (the average functionality is about 4, Table VIII). This suggests that the kinetic chain length of the copolymer is regulated by the number of the allylic monomers which enter into the copolymer chain. Because the chain propagation of the allylic monomer is so slow that it kinetically controls the growth of the copolymer chain, and, therefore, only the short chain or the low molecular weight copolymers can be formed.

Table VII. Dependencies of Molecular Weight and Functionality on Composition of Allyl Propoxylate and Methyl Methacrylate Copolymers*

Allyl Propoxylate/ MMA in mole	15:85	27:73	33:67	40:60
Mw	31080	13790	10470	7260
Mn	7300	4820	3990	3030
Mw/Mn	4.6	2.8	2.6	2.4
Number Average Functionality	11.7	12.5	12.4	11.0

* Polymers prepared at 135ºC

Table VIII. Dependencies of Molecular Weight and Functionality on Composition of Allyl Alcohol and Styrene Copolymers*

Allyl Alcohol/ Styrene in mole	7.5:92.5	16:84	22:78	30:70	35:65
Mw	16740	6450	4360	3150	3000
Mn	5030	2610	2000	1440	1360
Mw/Mn	3.3	2.5	2.2	2.2	2.2
Number Average Functionality	3.8	4.3	4.7	4.8	5.4

* Polymers prepared at 135ºC

Hydroxyl Sequence-Length Distribution. The hydroxyl group distribution along the polymer chain can be defined by the distributions of the various lengths of continuous hydroxyl functional monomeric units, that is the hydroxyl sequence-length distribution. The probabilities or mole fractions, $(N_1)_x$ and $(N_2)_x$, of forming methyl methacrylate and allyl alcohol sequences of length x are given by the equations 8 and 9. $(N_2)_x$ represents the hydroxyl sequence-length distribution, that is, the sequence distribution of allyl alcohol monomeric unit.

$$(\underline{N}_1)_x = (p_{11})^{(x-1)} p_{12} \qquad (8)$$

$$(\underline{N}_2)_x = (p_{22})^{(x-1)} p_{21} \qquad (9)$$

where the p values are defined by

$$p_{11} = r_1 / \{r_1 + ([M_2]/[M_1])\}, \quad p_{12} = 1 - p_{11} \qquad (10)$$

$$p_{22} = r_2 / \{r_2 + ([M_1]/[M_2])\}, \quad p_{21} = 1 - p_{22} \qquad (11)$$

Listed in Table IX are the allyl alcohol sequence distributions of the copolymers of allyl alcohol with methyl methacrylate for various comonomer feed compositions. It shows that the dominant sequence in the copolymer is the mono sequence ($x=1$) in a broad range of the monomer feed compositions. For comparison, a copolymer of methyl methacrylate (M_1) and hydroxyethyl methacrylate (M_2) with $r_1 \approx r_2 \approx 1$ gives 50% of mono sequence and 50% of diad or longer sequences of the hydroxyl functional monomeric unit for an equimolar feed composition.

Table IX. Hydroxyl Sequence-Length Distributions
For MMA and AA Copolymerization @ 135oC
With Various Monomer Feed Compositions

Monomer Feed [MMA]/[AA]	Mono $x = 1$	Diad $x = 2$	Triad $x = 3$	Tetrad $x = 4$
1.0	98.1%	1.92%	0.038%	0.000%
0.75	97.4%	2.53%	0.066%	0.002%
0.50	96.2%	3.70%	0.143%	0.005%
0.25	92.6%	6.85%	0.507%	0.037%
0.10	83.3%	13.9%	2.32%	0.386%

Lactone Formation Along Polymer Chain. Six-member lactone has been identified in the copolymers of allyl alcohol and methyl methacrylate by ^{13}C- NMR (Fig. 1). The lactone is formed along the polymer backbone by trans-esterification of the methyl methacrylate monomeric units with adjacent allyl alcohol monomeric units (Equation 8). The concentration of the lactone is dependent on the copolymer composition, ie., the ratio of allyl alcohol and methyl methacrylate as showed in Table X. The lactone concentration reaches the maximum level when the molar ratio of allyl alcohol to methyl methacrylate in the copolymer is close to 1:1. A small

122

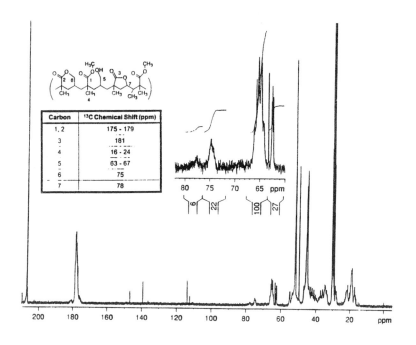

$$-\text{MeOH} \qquad\qquad\qquad (9)$$

amount of five-member lactone has also been found. It is formed from the vinyl isomer of allyl alcohol.

Table X. Dependence of Lactone Concentration On Copolymer Composition

MMA/AA Mole Ratio	Mole% of Lactone
1.2 : 1.0	12.3
2.0 : 1.0	7.8
4.5 : 1.0	4.0

Carbon	13C Chemical Shift (ppm)
1, 2	175 - 179
3	181
4	16 - 24
5	63 - 67
6	75
7	78

Fig 1. 13C-NMR Spectrum of the Copolymer of
Allyl Alcohol with Methyl Methacrylate

Evaluation of the Copolymer in Automotive Clearcoats

Hydroxyl functional acrylic resins for coatings have been developed by the bulk free radical polymerization of the allylic alcohol, including allyl alcohol and allyl propoxylate, with the acrylate and styrene comonomers [16-18]. The hydroxyl acrylic resins usually require more than one comonomer. The commonly used comonomers include methyl methacrylate, butyl methacrylate, butyl acrylate, and styrene. Each comonomer contributes different properties to the resin and its derivative coatings. Table XI lists three experimental hydroxyl acrylic resins which were prepared from allyl alcohol, allyl monopropoxylate and allyl propoxylate with average 1.6 propoxylate units (AAP1.6). These resins have been evaluated in the automotive clearcoat applications. The coatings formulated from the allylic alcohol based hydroxyl acrylic resins match the existing commercial systems based on hydroxyalkyl acrylate in performance including the film mechanical properties and the environmental resistances (Table XII). The non-solvent polymerization process based on the allylic alcohol offers advantages over the existing technology in increasing production efficiency and lowering the process cost. The non-solvent process is particularly important in producing powder and water borne coatings because the process solvent has to be removed from the final product. The inherent characteristics

Table XI Compositions and Physical Properties of Allylic Alcohol Based Hydroxyl Acrylic Resins

Copolymer Composition	Resin A-120	Resin P-120	Resin AP-120
Allyl Alcohol	12	-	-
Allyl Monopropoxylate	-	27	-
AAP 1.6	-	-	35
Styrene	8	7	9
Methyl Methacrylate	8	7	6
Butyl Methacrylate	60	50	44
Butyl Acrylate	12	9	6
Total	**100**	**100**	**100**
Copolymer Physical Properties			
Mw, GPC	6300	10800	10700
Mn, GPC	2200	3300	3100
OH#, mgKOH/g	125	130	128
Solvent	PMAc*	PMAc	PMAc
Solid %	70	70	70
Viscosity, Brookfield @25°C	3,800	715	588
Tg, °C, calculated	20	10	2

* Propylene glycol methyl ether acetate

of the allylic alcohol based hydroxyl acrylic resins in the even hydroxyl group distribution and the consistent hydroxyl functionality are under further evaluation in developing new coatings which are of reduced VOC (Volatile Organic Compound) and improved performance.

Table XII   Automotive Clearcoat Evaluations of Allyl Alcohol Based Hydroxyl Functional Acrylic Resin, A-120

Film Properties	Test Method	Control**	A-120 Based
Solvent Resistance	GM9509P	0	0
Chip Resistance	GM9508P	7	7
Gasoline Dip	GM9501P	20th Cycle	20th Cycle
Adhesion Method B	GM9071P	99%	100%
Quick Knife Adhesion	GM9552P	No Peeling	No Peeling
Koing Hardness	ASTM D4366	60	56
Impact Resistance, kg/m		2.86×10^3	2.86×10^3
Humidity	GM4465P	No Change	No Change
Environmental Damage Resistance		-	Equal to Control
Gloss Retention @ 20o			
4000 hours Xenon Arc	SAE J1960	65%	67%
24 months Florida Exposure	GM916JP	91%	90%

* A 2K polyurethane Clearcoat
** A high performance commercial 2K polyurethane clearcoat based on a HEA hydroxyl acrylic resin

EXPERIMENTAL

Procedure For Preparing Allyl Alcohol And Acrylate Copolymer

The preparation was carried out in a stainless steel pressure vessel in a hood with all due precautions, particularly in regard to toxicity and fire hazards related to handling allyl alcohol , acrylate monomers, and peroxide initiator. n-Butyl acrylate (204 grams), n-butyl methacrylate (1064 grams), methyl methacrylate (143 grams), and styrene (143 grams) were mixed. Allyl alcohol (437 grams), part of the comonomer mixture (204 grams), and part of di-t-butyl peroxide (18 grams) were charged into the reactor, and then the reactor was purged three times with nitrogen, and sealed. The remaining comonomer mixture and initiator were charged into the monomer addition pump and the initiator addition pumps, respectively. The reactor contents were heated to 135ºC in less then 30 minutes, and the comonomer mixture and the initiator in the addition pumps were gradually added into the reactor during the polymerization in the following decreasing addition rates:

Time at 135ºC	Comonomer Addition	Initiator Addition
0-1 hour	400 grams	20 grams
1-2 hour	350 grams	18 grams
2-3 hour	300 grams	14 grams
3-4 hour	200 grams	9 grams
4-5 hour	100 grams	4 grams

The polymerization continued for 30 minutes after completing the comonomer and the initiator additions. The unreacted monomers were removed by distillation under vacuum by heating up to 160ºC for about 2 hours or till the allyl alcohol residue level below 50 ppm and the total monomer residue less than 500 ppm. The resulting hydroxyl acrylic resin was discharged into a container which contained the solvent desired for the end application.

Experimental Determination of Monomeric Reactivity Ratios
The monomeric reactivity ratios and the copolymerization rate constants were evaluated by determining the copolymer compositions for several different comonomer feed compositions under a low degree of monomer conversion (< 5%). The copolymerization was carried out in an one-liter pressure stainless steel reactor at 135ºC. The reactor was charged with 500 grams of allylic alcohol and the desired amount of comonomer. After the reactor content was heated to 135ºC, the initiator (di-t-butyl peroxide) was injected into the reactor with agitation. The copolymerization was quenched by fast cooling to 25ºC after 2-10 minutes of copolymerization. The unreacted monomers were removed by vacuum distillation, and the copolymer compositions were measured by ^{13}C-NMR.

Characterization of Lactone Structures
^{13}C-NMR spectra were acquired using a Bruker DMX spectrometer at a magnetic field strength of 7.05 T (1H frequency = 300 MHz). An inverse-gated 1H decoupled experiment was used with the WALTZ-16 composite-pulse decoupling sequence and a relaxation delay of 12 s. The sample was analyzed as a 50 % solution in acetone-d6 at a temperature of 27º C. Chemical shifts are referenced to the central methyl peak of acetone-d6 at 29.8 ppm.

CONCLUSION

Allylic alcohols are hydroxyl functional monomers which readily copolymerize with styrene and acrylate to produce valuable hydroxyl functional resins. Allylic alcohols not only provide the copolymers with hydroxyl functional groups but also control the molecular weight of the copolymers. The monomeric reactivity ratios of allylic alcohols are considerably lower than those of styrene or acrylate, and, therefore, an excess of allylic alcohol is used to achieve a high hydroxyl concentration of the copolymer. The bulk semi-batch process produces the copolymers of low molecular weight and narrow molecular weight distribution. The copolymers are uniquely featured by a consistent hydroxyl functionality among the polymer chains which are of different molecular weights, and by even hydroxyl sequence distribution. The copolymers provide more uniform crosslinked network than the conventional resins based hydroxyalkyl acrylate. They are of great potential in applications of coatings, elastomers, adhesives, inks, and many other thermoset polymer areas.

126

ACKNOWLEDGEMENT

This work was supported by AA Based Acrylic Polyols Project Team of ARCO Chemical Company. In particular, Mark Smithson, Robert Good and Julia Weathers conducted the laboratory work in preparing the polymer samples, and David Kinney provided NMR analysis.

REFERENCES

1   Swern, 1., J. Am. Chem. Soc., **1949**, 71, 1152.
2   Harris S., Good A., Good R., and Guo S. H., Modern Paint and Coatings, **1994**, November, 34.
3   Guo, S. H., U.S. Patent 5,444,141 (1995).
4   Guo, S. H., U.S. Patent 5,512,642 (1996).
5   Guo, S. H., U.S. Patent 5,475,073 (1995).
6   Culbertson, B. M., Tony, Y., and Wan, Q., J. Macromol. Sci.-Pure Appl. Chem., **1997**, A34(7), 1249.
7   Nakayama, Y., Watanabe, T., and Toyomoto, I., J. of Coatings Technology, **1984**, 56(716), 73.
8   Pourreau, D., Guo, S. H., and Corujo, B., J. American Paint & Coatings **1995**, May, 43.
9   Guo, S. H., and Gastinger, R., U.S. Patent 5,420,216 (1995).
10   Odian, G., Principles of Polymerization, John Wiley & Sons, New York, NY, 2nd Ed., **1981**, p450.
11   Odian, G., Principles of Polymerization, John Wiley & Sons, New York, NY, 2nd Ed., **1981**, p440-441.
12   Organic Peroxide, Product Bulletin, Atochem North America, Inc, Buffalo, NY.
13   Odian, G., Principles of Polymerization, John Wiley & Sons, New York, NY, 2nd Ed., **1981**, p258.
14   Litt, M.; and Eirich F. R., J. Polym. Sci., **1960**, 45, 379.
15   Ramby B., J. Appl. Polym. Sci. Appl. Polym. Symp., **1975**, 126, 327.
16   Guo, S. H., U.S. Patent 5,525,693 (1996).
17   Guo, S. H., U.S. Patent 5,646,213 (1997).
18   Guo, S. H., U.S. Patent 5,646,225 (1997).

Chapter 8

A Versatile Synthesis of Block and Graft Copolymers by Bulk 'Living' Free Radical Procedures

Craig J. Hawker[1], James L. Hedrick[1], Eva Malmström[1], Mikael Trollsås[1], Udo M. Stehling[2], and Robert M. Waymouth[2]

[1]Center for Polymeric Interfaces and Macromolecular Assemblies, IBM Almaden Research Center, 650 Harry Road, San Jose, CA 95120–6099
[2]Center for Polymeric Interfaces and Macromolecular Assemblies, Department of Chemistry, Stanford University, Stanford, CA 94305

The stability of alkoxyamine unimolecular initiators to a variety of reaction conditions, combined with the compatibility of nitroxide mediated 'living' free radical procedures with numerous functional groups, offers an unprecedented opportunity to prepare novel block and graft copolymer structures from a wide selection of different monomer units and chemistries. This unique feature is illustrated by three different approaches to block copolymers by solvent free 'living' radical polymerization; (i) controlled polymerization of macromonomers; (ii) formation of graft copolymers from multifunctional polymeric initiators; and (iii) synthesis of block copolymers by dual living polymerizations from a double headed initiator.

The development of a free radical polymerization which displays many of the characteristics of a living polymerization has long been the goal of polymer chemists *(1)*. The driving force for this interest is the ability of living systems to accurately control macromolecular structure and architecture. A major drawback of most living systems is however the extreme reactivity of the initiating and propagating species. This leads to incompatibility with a wide range of functional groups as well as very demanding reaction conditions and purification requirements. Therefore the development of a 'living' free radical procedure which combines the level of control associated with living systems, with the functional group tolerance of free radical chemistry, would be an extremely attractive methodology for synthetic polymer chemistry.

Early attempts to realize a "living", or controlled free radical process involved the concept of reversible termination of growing polymer chains by iniferters *(2,3)*. This concept was however plagued by high polydispersities and poor control over molecular weight and chain ends. Following this approach, Moad and Rizzardo introduced the use of stable nitroxide free radicals, such as 2,2,6,6-tetramethylpiperidinyloxy (TEMPO), as reversible terminating agents to "cap" the growing polymer chain *(4)*. In a seminal contribution, the use of TEMPO in 'living' free radical polymerizations was refined by Georges et al who demonstrated that at elevated temperatures narrow molecular weight distribution polymers (PD = 1.1-1.3) could be prepared using bulk polymerization conditions *(5)*.

The key features of nitroxide-mediated "living" free radical polymerizations are that the carbon-oxygen bond of the dormant, or inactive, alkoxyamine 1 is homolytically unstable and undergoes thermal fragmentation to give a stable nitroxide 2 and the polymeric radical, 3. Significantly, the nitroxide free radical 2 does not initiate the growth of any extra polymer chains, which is in direct contrast to iniferters, but it does react at near diffusion controlled rates with the carbon-centered free radical of the propagating polymer radical, 3. This allows controlled growth, or chain extension, to give a polymeric radical, 4, in which the degree of polymerization has increased. Recombination of 4 with the mediating nitroxide radical then gives the dormant species, 5, and the cycle of homolysis-monomer addition-recombination can be repeated (Scheme 1). The presence of significant amounts of covalent chain ends, such as 1 or 5, leads to a substantial decrease in the overall concentration of radical chain ends. An extremely favorable consequence of this is that the occurrence of unwanted side reactions such as combinations, disproportionation, or termination is decreased substantially which leads to controlled, or pseudo-living, growth. However it should be noted that the occurrence of these side reactions is not eliminated so, in the strictest sense, the polymerizations are not truly living, though they show many of the desirable characteristics of living polymerizations.

The scope of nitroxide mediated 'living' free radical polymerization has been subsequently been extended to produce narrow polydispersity materials with controlled molecular weights *(6)*, chain ends *(7)*, and chain architectures *(8)*. Also well-defined random *(9)* and block *(10)* copolymers can easily be prepared from a variety of monomers. The chemical stability of these alkoxyamine initiators under a variety of reaction conditions permits the introduction and manipulation of numerous functional groups. These functionalized initiators can then be used to prepare novel polymeric materials which are vital for probing a range of problems in material science *(11)*.

Architectural Control

Using the 'living' free radical polymerization procedure described above, coupled with the concept of alkoxyamine based unimolecular initiators permitted a thorough examination of the synthesis of block and graft copolymers using this novel technique. Traditionally these structures are constructed using synthetically demanding living procedures such as anionic or cationic techniques *(12,13)*. While successful, these techniques are not compatible with the formation of block or graft copolymers from reactive monomers/polymers or from dissimilar monomer systems which are polymerized by fundamentally different chemistries (i.e. anionic ring opening and free radical procedures). In this report we demonstrate the versatile synthesis of a variety of block and graft copolymers using functionalized unimolecular initiators under typical bulk 'living' free radical conditions. Three techniques will be described; 1) controlled polymerization of macromonomers; 2) growth of graft copolymers from a multifunctional polymeric initiator; and 3) formation of block copolymers from a dual, or difunctional, initiator.

Macromonomer Polymerization

The polymerization of macromonomers has been studied extensively and shown to be a useful technique for the synthesis of graft copolymers. One drawback to the currently available techniques is that the preparation of well defined graft copolymers from macromonomers containing either reactive polymeric linkages, or reactive functional groups along the macromonomer backbone, is extremely problematic. The use of such macromonomers in living anionic or cationic techniques can lead to cleavage or unwanted side reactions, while more benign polymerization techniques, such as condensation or normal free radical procedures, fail to give well defined materials.

Recently it has been shown that 'living' free radical techniques offer a viable alternative for the synthesis of a wide variety of well defined graft copolymers from reactive macromonomers which overcomes many of the above problems (14). For example, the polymerization of 5 equivalents of the methacrylate terminated poly(D,L)lactide (M_n = 2,300; PD. = 1.25), **6**, with 450 equivalents of styrene in the presence of the unimolecular initiator, **7**, was conducted under typical bulk polymerization conditions to give the desired graft copolymer, **8** (M_n = 45 000; PD. = 1.35). Interestingly, the solubility/compatibility of the poly(lactide) macromonomer, **6**, under the bulk reaction conditions did not lead to any complications and no precipitation, or phase separation, of the poly(lactide) was observed during the course of the polymerization. Purification by precipitation from tetrahydrofuran into methanol removed trace amounts of the unreacted macromonomers/monomers and afforded the graft copolymer, **8**, as shown in Scheme 2. Continued extraction of **8** with refluxing methanol did not lead to any further change in the composition of the graft copolymer and the incorporation of the poly(lactide) graft into the polystyrene backbone was demonstrated by a number of spectroscopic and chromatographic techniques (Figure 1). It was subsequently shown for a variety of different macromonomer systems, ranging from polyesters to polyethylene glycols, that by simply changing the ratio of monomers to unimolecular initiator, **7**, the molecular weight of the final graft copolymer could be accurately controlled while maintaining low polydispersities. Also the weight percent of macromonomer could be increased to ca. 40% under bulk polymerization conditions without any complications, though it should be noted that above ca. 50 wt% solubility of a number of macromonomers in the polymerization mixture became problematic. Addition of solvents to these high loading systems overcame the solubility problems but at very high loadings (ca. 75 wt%) dilution of the vinyl groups became a problem and the polymerizations slowed down considerably which agrees with earlier studies on the effect of solvent on nitroxide mediated 'living' free radical polymerizations (15).

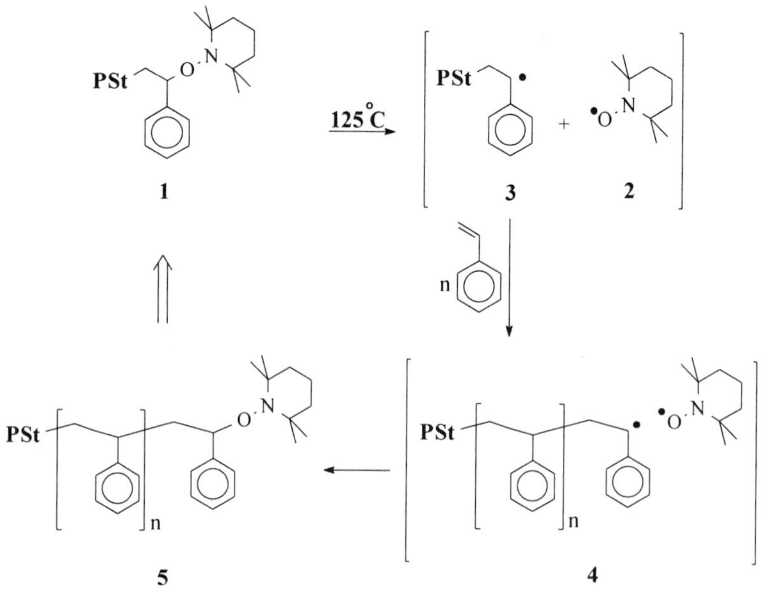

Scheme 1

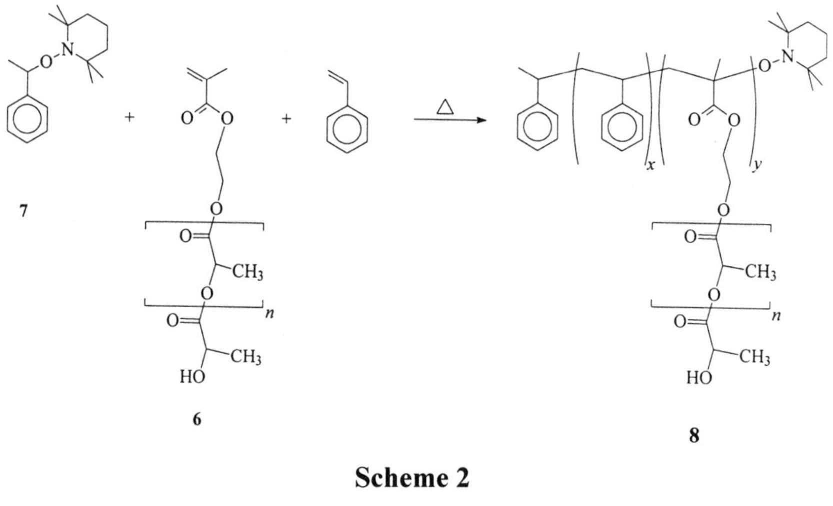

Scheme 2

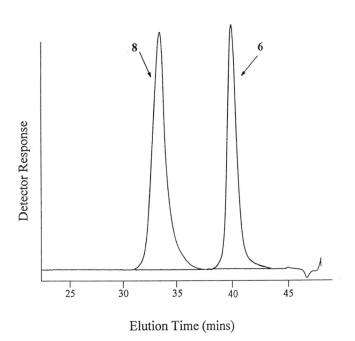

Elution Time (mins)

Figure 1. Comparison of GPC traces for the poly(lactide) macromonomer, **6**, and the graft copolymers, **8**, prepared by 'living' free radical polymerization.

Multifunctional Polymeric Initiators

One of the unique features of alkoxyamine based unimolecular initiators is their compatibility with a wide range of reaction conditions. This not only permits a variety of functionalized alkoxyamines to be prepared, but also allows a number of polymerization reactions to be performed in their presence with no loss of initiating ability. The opportunity therefore exists to prepare novel multifunctional polymeric initiators which can lead to graft copolymers. Initially, we demonstrated the feasibility of this process by preparing the styrene-alkoxyamine adduct, **9**, by reaction of the hydroxy derivative, **10**, with p-chloromethylstyrene in the presence of sodium hydride *(16)*. Copolymerization of **9** with styrene was the conducted under normal free radical polymerization conditions using AIBN as an initiator in refluxing tetrahydrofuran. Significantly, the polymerization proceeded smoothly to give the desired copolymer **11** (M_n = 12 000; PD. = 1.80) in 78% yield after purification (Scheme 3). Analysis of the copolymer by ^{1}H and ^{13}C NMR spectroscopy showed the expected resonances for the alkoxyamine group and comparison with the aromatic styrenic resonances for the backbone polymer permitted the ratio of monomer units to be determined. The experimentally determined value of 1:19 agrees with the feed ratio and shows that the alkoxyamine group is stable to normal free radical conditions and has little, if any, effect on reactivity ratios.

Another unique feature of 'living' free radical polymerizations which is crucial to the success of this multifunctional polymeric initiator approach is the low occurrence of radical-radical coupling reactions. This allows a number of vinyl polymer chains to be grown simultaneously from a single backbone polymer without complications from gelation and chain-chain coupling which would occur under normal free radical conditions. To demonstrate this approach, bulk polymerization of a mixture of the copolymer, **11**, and 200 equivalents of styrene at 130°C for 72 hours resulted in polymerization of the added styrene to give the proposed graft system, **12**. Comparison of the GPC traces for the starting polymer, **11**, and the graft system, **12**, clearly shows an increase in molecular weight for **12** and the absence of unreacted starting polymer. In this case the nature of the grafted polymer chain, **13**, could not be probed by hydrolysis due to the stability of the ether linkage and cleavage of this group was therefore accomplished by treatment of **12** with an excess of trimethylsilyl iodide. This results in a dramatic change in the GPC trace for the isolated product, no peak for the starting graft system is observed and the polydispersity of the sample is lowered from 2.01 to 1.26. Interestingly, the number average molecular weight for **13** was determined to be 23 000 which agrees closely with the theoretical M_n for the grafted chains of 21 000. These results demonstrate that a graft system is produced and that the TEMPO groups attached to the polystyrene backbone of **11** are capable of initiating the polymerization of styrene to give grafts of controlled molecular weight and low polydispersity.

Subsequently, we have extended this approach to the growth of functionalized random copolymer brushes which are useful for the construction of combburst-type macromolecules *(17)* and to the synthesis of novel poly(olefin) graft copolymers *(18)*.

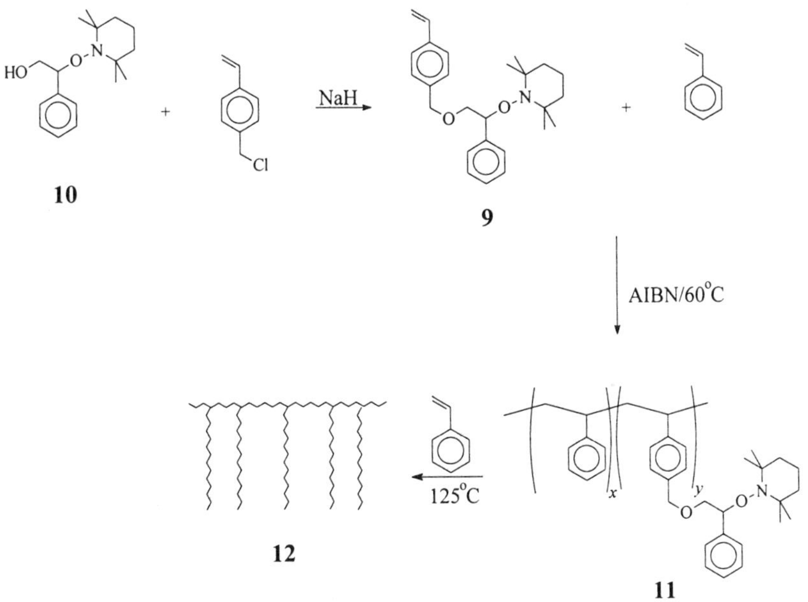

Scheme 3

The latter materials are prepared by the metathesis polymerization of the alkene functionalized alkoxyamine, 14, to give the functionalized poly(olefin), 15, which can then be used to prepare a variety of well defined graft copolymers such as 16 (Scheme 4).

Dual Initiating Systems

This compatibility of alkoxyamine groups with a variety of reaction conditions was also exploited in the design of a simple and versatile synthesis of AB diblock copolymers by dual initiation from a double headed initiator. In this novel approach the initiator, 10, is designed to have two different, though compatible, initiating sites from which two polymer chains can be grown sequentially without the need for any intermediate functionalization or activation steps. For example, the hydroxy functionalized alkoxyamine, 10, contains a single primary alcohol which can be used as the initiating center for the living ring opening polymerization of cyclic lactones, as well as a secondary benzylic group which is an efficient initiator for the nitroxide mediated 'living' free radical polymerization of vinyl monomers. To demonstrate this principle it was initially decided to grow a well defined polycaprolactone chain from the hydroxy group of 10 and then, without further chemical transformations, use the alkoxyamine functionalized polycaprolactone as a monofunctionalized telechelic initiator for the controlled polymerization of vinyl monomers leading to a highly efficient synthesis of AB block copolymers which are difficult to prepare by other means. The living ring opening polymerization of ε-caprolactone, 17, by 10 as the initiator was studied using a catalytic amount of aluminum tris(isopropoxide) as a promoter and the compatibility of the alkoxyamine group with the polymerization conditions was evidenced by the fact that the polycaprolactones, 18, obtained using this procedure had extremely low polydispersities and controllable molecular weights (19). For example, polymerization of 70 equivalents of 17 by 10 gave polycaprolactone 18 with a polydispersity of 1.04 and a degree of polymerization of 74 as determined by ^{1}H NMR spectroscopy. More importantly, resonances for the single alkoxyamine group at the chain end could be observed at 0.6, 4.2-4.8, and 7.27 ppm.

The functionalized polycaprolactone, 18, could then be used to initiate the polymerization of vinyl monomers, such as styrene or a mixture of styrene and methyl methacrylate, to give a block copolymer, 19, with no intermediate activation or functionalization steps. As shown in figure 2, the molecular weight of the block copolymer increased in a systematic way with increasing amounts of styrene and GPC analysis could not detect any unreacted polycaprolactone in the block copolymers. For example, polymerization of 275 equivalents of styrene with polycaprolactone, 18, having a molecular weight of 8 500 gave the block copolymer, 19, which was shown to have a number average molecular weight of 37 000 (theoretical molecular weight of 34 000) and a polydispersity of 1.09 (Scheme 5). Significantly, the method of construction could be easily reversed without any change in the degree of control over AB block copolymer formation. Therefore the vinyl polymer block can be grown

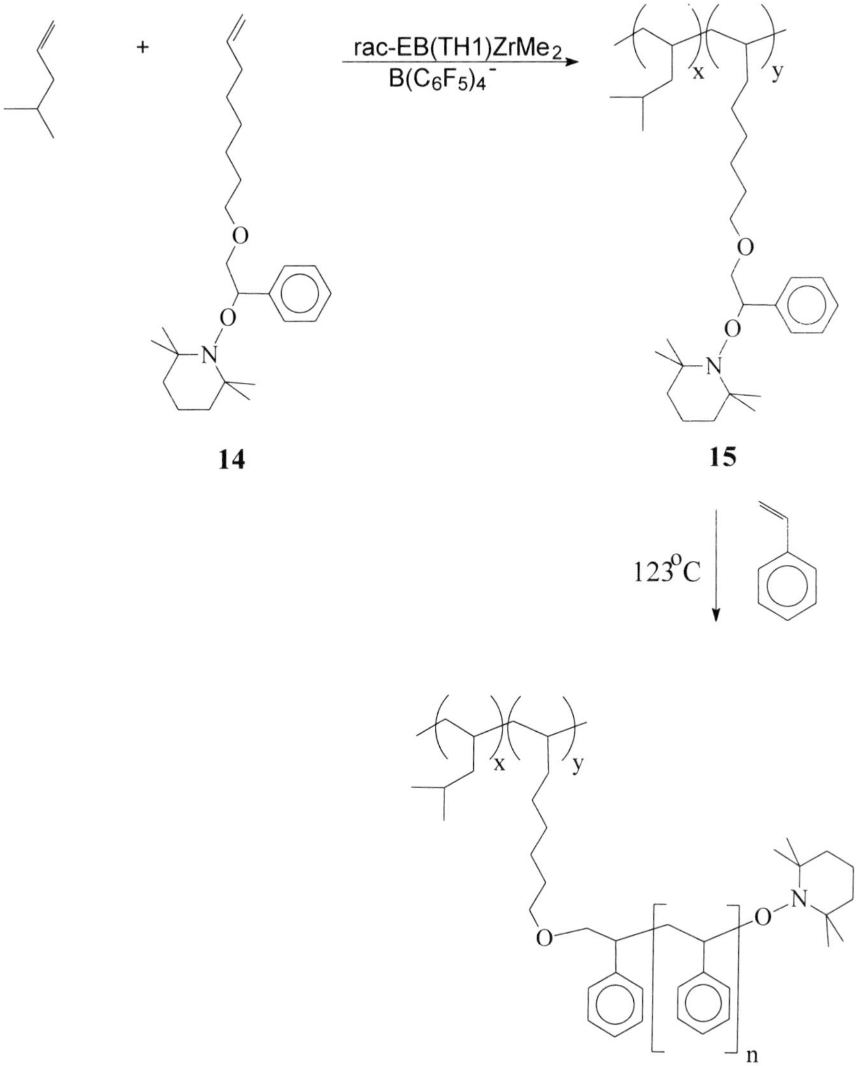

Scheme 4

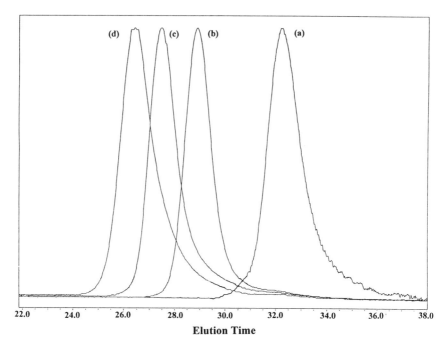

Figure 2. Comparison of GPC traces for (a) the functionalized poly(caprolactone) initiator, **18**, and the block copolymers, **19**, obtained from the reaction with varying amounts of styrene; (b) 250:1, (c) 600:1, and (d) 800:1.

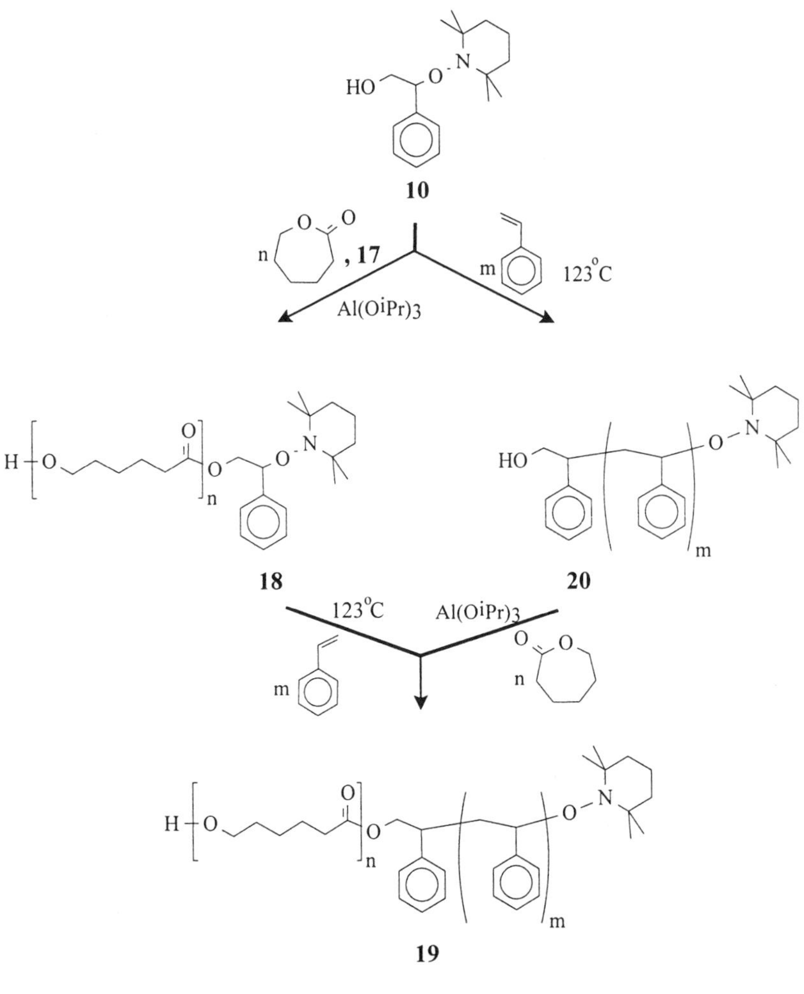

Scheme 5

from the alkoxyamine initiating center to give the hydroxy-terminated macromolecule, **20**, which can be used as an initiator for the ring-opening polymerization of caprolactone to give the equivalent block copolymer, **19**, as prepared above. It should be noted that in both cases the degree of polymerization of each block can be accurately controlled by the molar ratio of monomer to initiator.

Conclusion

In conclusion, we have demonstrated the compatibility of alkoxyamine initiating centers with a variety of polymerization conditions. These mono- or multi-functional polymeric initiators can then be used under typical bulk polymerization conditions to give novel block and graft copolymers. This permits novel well defined block copolymers to be readily prepared in the minimum number of steps under synthetically non-demanding conditions.

Acknowledgments: The authors gratefully acknowledge financial support from the NSF funded Center for Polymeric Interfaces and Macromolecular Assemblies and the IBM Corporation. Fellowship support from the Hans Werthén Foundation (EM), Sweden, the Swedish Institute (MT), the Fulbright Commission (EM, MT), and BASF AG (US) are also acknowledged.

References
(1) Webster, O. W. *Science* **1994**, *251*, 887; Fréchet, J.M.J. *Science* **1994**, *263*, 1710.
(2) Otsu, T.; Matsunaga, T.; Kuriyama, A.; Yoshida, M. *Eur. Polym. J.* **1994**, *25*, 643.
(3) Turner, S.R.; Blevins, R.W. *Macromolecules* **1990**, *23*, 1856.
(4) Moad, G.; Solomon, D.H.; Johns, S.R.; Willing, R.I. *Macromolecules* **1982**, *15*, 1188.
(5) Georges, M.K.; Veregin, R.P.N.; Kazmaier, P.M.; Hamer, G.K. *Macromolecules* **1993**, *26*, 2987.
(6) Hawker, C.J. *J. Am. Chem. Soc.*, **1994**, *116*, 11314; Hammouch, S.O; Catala, J.M. *Macromol. Rapid Commun.* **1996**, *17*, 149.
(7) Hawker, C.J.; Hedrick, J.L. *Macromolecules*, **1995**, *28*, 2993; Catala, J.M.; Bubel, F.; Hammouch, S.O. *Macromolecules* **1995**, *28*, 8441; Frank, B.; Gast, A.P.; Russell, T.P.; Brown, H.R.; Hawker, C.J. *Macromolecules* **1996**, *29*, 6531.
(8) Hawker, C.J. *Angew Chem. Int. Ed. Engl.*, **1995**, *34*, 1456; Gaynor, S.; Edelman, S.; Matyjaszewski, K. *Macromolecules*, **1995**, *28*, 6381; Hawker, C.J.; Mecerreyes, D.; Elce, E.; Dao, J.; Hedrick, J.L.; Barakat, I.; Dubois, P.; Jerome, R.; Volksen, W. *Macromol. Chem. Phys.*, **1997**, *198*, 155; Hawker, C.J.; Fréchet, J.M.J.; Grubbs, R.B.; Dao J. *J. Am. Chem. Soc.* **1995**, *117*, 10763.
(9) Hawker, C.J.; Elce, E.; Dao, J.; Russell, T.P.; Volksen, W.; Barclay, G.G. *Macromolecules* **1996**, *29*, 2686.

(10)   Fukuda, T.; Terauchi, T.; Goto, A.; Ysujii, Y.; Miyamoto, T.; Shimizu, Y. *Macromolecules*, **1996**, *29*, 3050; Kobatake, S.; Harwood, H.J.; Quirk, R.P.; Priddy, D.B. *Macromolecules*, **1997**, *29*, 4238; Bertin, D.; Boutevin, B. *Polym. Bull.*, **1996**, *37*, 337; Hawker, C.J. *Acc. Chem. Res.*, **1997**, *30*, 373.

(11)   Mansky, P.; Liu, Y.; Huang, E.; Russell, T. P.; Hawker, C. J. *Science* **1997**, *275*, 1458; Kulasekere, R.; Kaiser, H.; Ankner, J. F.; Russell, T. P.; Brown, H. R.; Hawker, C. J.; Mayes, A. M. *Macromolecules* **1996**, *29*, 5493; Kulasekere, R.; Kaiser, H.; Ankner, J. F.; Russell, T. P.; Brown, H. R.; Hawker, C. J.; Mayes, A. M. *Physica B* **1996**, *221*, 306.

(12)   Quirk, R.P.; Kinning, D.J.; Fetters, L.J. *Comprehensive Polymer Science, Vol. 7*, Pergamon Press, London, Aggarwal, S.L. Editor, **1989**, 1.

(13)   Fradet, A. *Comprehensive Polymer Science, 2nd Supp.*, Pergamon Press, London, Aggarwal, S.L.; Russo, S. Eds, **1996**, 133.

(14)   Hawker, C.J.; Mecerreyes, D.; Elce, E.; Dao, J.; Hedrick, J.L.; Barakat, I.; Dubois, P.; Jerome, R.; Volksen, W. *Macromol. Chem. Phys.*, **1997**, *198*, 155.

(15)   Hawker, C.J.; Barclay, G.G.; Orellana, A.; Dao, J.; Devonport, W. *Macromolecules* **1996**, *29*, 5245.

(16)   Hawker, C.J. *Angew Chem. Int. Ed. Engl.*, **1995**, *34*, 1456.

(17)   Grubbs, R.B.; Hawker, C.J.; Dao J.; Fréchet, J.M.J. *Angew. Chem. Int. Ed. Eng.* **1997**, *36*, 270.

(18)   Hawker, C.J.; Hedrick, J.L.; Malmstrom, E.E.; Trollsas, M.; Waymouth, R.M.; Stehling, U.M. *Polym. Prep.*, **1997**, *38(2)*, 412.

(19)   Hawker, C.J.; Hedrick, J.L.; Malmström, E.E.; Trollsås, M.; Mecerreyes, D.; Moineau, G.; Dubois, Ph.; Jêrome, R. *Macromolecules,* **in press**.

Chapter 9

Solvent-Free Synthesis by Free-Radical Frontal Polymerization

John A. Pojman[1], Dionne Fortenberry[1], Lydia Lee Lewis[1], Chris Simmons[1], and Victor Ilyashenko[2]

[1]Department of Chemistry and Biochemistry, University of Southern Mississippi, Hattiesburg, MS 39406-5043
[2]Boston Optical Fiber Company, 155 Flanders Road, Westborough, MA 01581

Frontal polymerization is a mode of converting monomer into polymer via a localized propagating reaction zone. Such fronts can exist with free-radical polymerization and epoxy curing, and they allow the safe direct conversion of monomer into polymer without the use of solvent. The conditions for the existence of the free-radical thermal frontal polymerization and for isothermal frontal polymerization are considered. The factors affecting velocity, conversion, and molecular weight for thermal frontal polymerization are also considered. Both thermal and convective instabilities are examined as well as strategies for their elimination.

In 1972 at the Institute of Chemical Physics in Chernogolovka (Russia), Chechilo and Enikolopyan discovered thermal frontal free-radical polymerization of vinyl monomers. Using a metal reactor under high pressure (> 3000 atm) they studied descending fronts of methyl methacrylate with peroxide initiators (1,2). In 1991 Pojman rediscovered this phenomenon using methacrylic acid at ambient pressure in standard glass test tubes (Figure 1) (3).

In this chapter, we will provide an overview of thermal fronts followed by a similar discourse on isothermal fronts. Types of systems that exhibit thermal frontal polymerization, basic aspects of frontal polymerization, the nature of the product produced, including conversion and molecular weight distribution, and both convective and thermal instabilities will also be discussed.

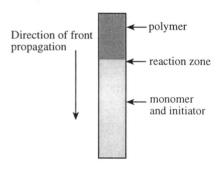

Figure 1. A schematic diagram of frontal polymerization

I. Basic Phenomena in frontal polymerization

1. Types of Systems In the early work of frontal polymerization, the experimenters ran fronts at extremely high pressures (up to 5000 atm) to prevent monomer boiling and to eliminate the Rayleigh-Taylor (see below) instability formed by density gradients in the reaction zone (*1,2,4,5*). They were also only able to run descending fronts because buoyancy-driven convection would remove enough heat to extinguish an ascending front. At lower pressures (1500 atm) the descending fronts were destroyed because of the density gradient that formed as the reaction progressed.

Unfortunately, we do not know the necessary and sufficient conditions to sustain a thermal front. However, we do know several factors that favor this thermal frontal mode. The monomer must have a boiling point above the front temperature to prevent heat loss from vaporization. (The boiling point can be raised by applying pressure.) Bubbles formed from this vaporization can also obstruct the transfer of heat. Because the rate of heat production must exceed the rate of heat loss, a highly exothermic reaction is favorable. The dimensions of the reactor or test tube can also play a significant role. If the surface area to volume ratio is too large, even a reactive system will be quenched. For example, at room temperature the only system we have found that can propagate in a 3 mm glass tube is acrylamide -- no liquid monomer is sufficiently reactive or exothermic. In light of these facts, we will summarize our experimental results (*3,6-8*) and reliable experimental data from other authors (*1,2,4,5,9*) to outline the conditions under which frontal polymerization is expected.

The first case is multifunctional monomers that undergo crosslinking including, but not limited to, tri(ethylene glycol)dimethacrylate (TGDMA), 1,6 hexanediol diacrylate, and divinylbenzene. The free-radical polymerization of these monomers produces a rigid crosslinked polymer, which sustains a sharp frontal interface. Monofunctional monomers that are crosslinked with a compatible multifunctional monomer also fall into this category.

The second class of monomers that can sustain fronts forms polymer that is insoluble in its corresponding monomer, such as acrylic and methacrylic acids (*3,6,7*). The insoluble polymer particles adhere to each other and to the reactor or glass tube. The adhesive forces overcome the buoyancy-driven convection resulting from the density gradient at the polymer/monomer interface. This prevents collapse of the front, although Rayleigh-Taylor instability does still manifest partially in the form of fingering (see Figure 2) (*3,10*). How well the front sustains itself depends on conversion, the polymer glass transition temperature and molecular weight distribution. Indeed, these properties themselves depend on the initial reactant temperature, initiator type and concentration (*7*).

By rotating a tube around the axis of front propagation, Nagy and Pojman were able to suppress fingers formed in a methacrylic acid front (*11*). The front velocity depended on the fourth power of the rotational frequency, and the amplitude of the front curvature was proportional to the square of the frequency.

The third group of monomers includes all highly reactive monomers that produce thermoplastics that are molten at the front temperature. Such fronts decay due to the Rayleigh-Taylor instability (Figure 3). Although these polymers are soluble in their monomers (unlike the second class), on the time scale of the front, the polymer is effectively immiscible with the monomer. Adding an inert filler such as ultra-fine silica gel (CAB-O-SIL) or a soluble polymer, increases the viscosity and eliminates the front collapse. Some monomers like styrene and methyl methacrylate require moderate pressure (20-30 atm) to eliminate monomer boiling. Higher boiling temperature monomers like benzyl acrylate support the frontal regime at ambient pressure in test tubes. CAB-O-SIL is the preferred method for increasing viscosity because adding a small amount (1%-5% w/w) can change the viscosity over two to three orders of magnitude. It is not possible to add soluble polymer to obtain a similar viscosity and still have sufficient exothermicity to sustain a front. Frontal polymerization of the third group of monomers can be obtained in any orientation because the large viscosity (of the monomer/CAB-O-SIL system) suppresses buoyancy-driven convection.

The final class of monomers that can be frontally polymerized is solid monomers whose melting point is less than the adiabatic reaction temperature. One such monomer is acrylamide (13,14).

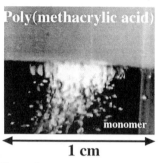

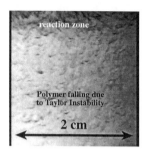

Figure 2   Figure 3

Figure 2. Descending "fingers" from a front of poly(methacrylic acid) polymerization. (Adapted from Nagy and Pojman) (11)

Figure 3. Rayleigh-Taylor instability in a descending front of n-butyl acrylate polymerization. (Adapted from Pojman et al.) (12)

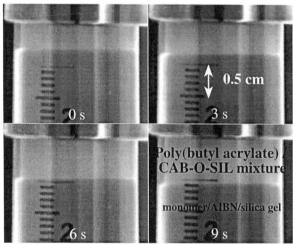

Figure 4. A montage of an n-butyl acrylate front propagating under 50 atm pressure. [AIBN] = 4% w/v. (Adapted from Khan and Pojman) (12)

2. Experimental Experiments for thermal frontal polymerizations are as follows. Monomer is added to a glass tube with initiator. For initiators, we use peroxides and nitriles predominately, although much work has been done with persulfate initiators as well. Although we typically use acrylate and methacrylate monomers, any number of different monomers can be used. In general, high boiling point and very reactive monomers are preferred. For these reasons, a typical monomer that we use is benzyl acrylate. Monomers whose boiling points are significantly lower than the front temperature must be polymerized under pressure. Figure 4 shows n-butyl acrylate polymerization. In the absence of any kind of thermal or convective instabilities the front moves with a constant velocity determined by a plot of the front position as a function of time (Figure 5).

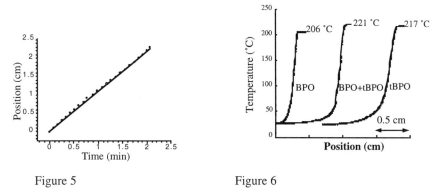

Figure 5   Figure 6

Figure 5. The front velocity (1.09 cm/min) is determined from the slope of a plot of front position versus time. This plot is for n-butyl acrylate polymerization under pressure. $[AIBN]_0 = 0.00457$ molal; initial temperature = 18.9 °C (Adapted from Pojman and Khan) (*15*)

Figure 6. The temperature profiles of methacrylic acid polymerization fronts with BPO alone, with benzoyl peroxide and t-butyl peroxide (tBPO) and with t-butyl peroxide alone. [BPO] = [tBPO] = 0.0825 mol/kg. Reactions were performed in 1.5 cm (i.d.) tubes. The temperatures indicated correspond to the maximum temperatures reached for each initiator. While the temperature at the monomer/polymer interface was not directly measured, it is believed to be *ca.* 100 °C (Adapted from Pojman et al.) (*7*)

3. Mechanism of Polymerization   A free-radical polymerization initiated thermally can be approximated by a three-step mechanism. First, the initiator must decompose to produce radicals:

$$I \longrightarrow f \, 2R\bullet \qquad (1)$$

where f is the efficiency, which is initiator dependent. A radical can then initiate a growing polymer chain:

$$R\bullet + M \longrightarrow P_1 \bullet \qquad (2)$$

$$P_n^{\bullet} + M = P_{n+1}^{\bullet} \qquad (3)$$

Step 3 continues until termination occurs by either combination or disproportionation.

Although the major heat release in the polymerization reaction occurs in the propagation step, it does not have a sufficiently high energy of activation to provide a frontal regime. Frontal polymerization autocatalysis occurs in the initiator decomposition step because the initiator radical concentration is the main control for the total polymerization rate, compared to the gel effect or direct thermal polymerization that may also be present in the frontal polymerization process. The steady-state assumption in the polymerization model gives an approximate relationship between the effective activation energy of the entire polymerization process and activation energy of the initiator decomposition reaction:

$$E_{eff} = E_p + (E_i / 2) - (E_t / 2) \qquad (4)$$

where E_p is the activation energy of the propagation step, E_i is for the initiator decomposition and E_t is that for the termination step.

The energy of activation of the initiator (E_i) has the largest magnitude in equation (4). Because of this, the initiator type plays a significant role in obtaining a front. It also determines the temperature profile and the velocity of the front. The exception here is 1, 6-hexanediol diacrylate, which is so highly reactive that it can be polymerized in the frontal mode without added initiator.

4. Temperature Profiles A thermal front has a sharp temperature profile (7), which can provide much useful information. From Figure 6 it may appear that the chemical reaction is occurring in a zone about 0.5 cm in width, which is not the case. If the chemical reaction has an infinitely high activation energy, the chemical reaction will occur in an infinitely narrow region. In actuality, the 0.5 cm represents a pre-heat zone. The temperature below that at which significant chemical reaction occurs follows an exponential profile. This can be described with the following relationship in terms of the front velocity (c), thermal diffusivity (κ) and temperature difference (ΔT) (16):

$$T(x) = T_0 + \Delta T \exp\left(\frac{cx}{\kappa}\right) \qquad (5)$$

Conversion is approximately proportional to the difference between the maximum and initial temperatures, and a higher maximum temperature indicates higher conversion. More stable initiators give higher conversion, as can be seen in Figure 4 with benzoyl peroxide (BPO) and t-butyl peroxide (tBPO). The methacrylic acid front with tBPO was significantly slower in spite of having the highest reaction temperature. This means that the effective activation energy of a polymerization front is directly correlated to the activation energy of the initiator decomposition.

5. Front Velocity Dependence on Initiator Concentration Chechilo et al. studied frontal polymerization of methyl methacrylate initiated with benzoyl peroxide. By using several thermocouples along the reactor at known positions, they could determine front velocity and found a 0.36 power dependence for the velocity on the benzoyl peroxide concentration (4). Pojman et al. did a similar study using TGDMA and different initiators finding a different power functional dependence for the three initiators AIBN, BPO, and LPO (7). Mathematical modeling of this dependence has been done by Goldfeder et al (17).

Pojman et al. studied binary systems with two non-interfering polymerization mechanisms (18). With a aliphatic amine cured epoxy and a free-radical cured diacrylate, they observed a minimum in the front velocity as a function of the relative concentration of each component. A comparable study is underway with copolymerization fronts.

II. Properties of Polymers

1. Conversion A problem with thermal propagating fronts is that the rapid temperature increase causes rapid initiator decomposition or "burnout", which can lead to low conversion (19). To have practical utility, higher conversion must be reached. One plausible solution is to use two initiators with drastically different activation energies, such that the more stable initiator will not decompose until the less stable initiator is consumed. Pojman et al. measured the conversion of double bonds as a function of initiator concentration (7) of such a system using benzoyl peroxide (BPO) and the more stable t-butyl peroxide (tBPO). These experiments showed that using a stable initiator provides better conversion, because the tBPO alone was almost as good as the dual system. (See Figure 5) The advantage of the dual system is that the conversion is determined by the more stable initiator but the front velocity is determined by the less stable one.

Conversion can also be thermodynamically limited. Because the polymerization reactions are exothermic, the equilibrium conversion decreases with increasing temperature (20). A relationship between temperature and the equilibrium monomer concentration (assuming unit activity coefficients) can be derived, in which $[M]_0$ is the standard monomer concentration used to calculate the ΔS^0 and ΔH^0.

$$T = \frac{\Delta H^o}{\Delta S^0 + R\ln([M]_{eq} / [M]_0)} \qquad (6)$$

For an adiabatic polymerization, the maximum conversion is uniquely determined by the ΔH^0 and ΔS^0 of polymerization. As the temperature increases, the equilibrium conversion is reduced and can be related by:

$$\alpha = 1 - \frac{1}{[M]_{initial}} e^{\dfrac{\Delta H^0 - T\Delta S^0}{RT}} \qquad (7)$$

The relationship for the temperature and conversion for adiabatic self-heating is:

$$T = T_i + \alpha\Delta H^0 / C_p \qquad (8)$$

The solution of Equations (8) and (9) provides the conversion achieved in adiabatic polymerization. Using thermodynamic data from Odian (21), a conversion of 0.93 is obtained for the adiabatic polymerization of methyl methacrylate with an initial temperature of 25 °C. This means that independent of initiator burnout, complete conversion can never be achieved because of the high front temperature. This value depends heavily on the exact values of the thermodynamic parameters so the calculated value may not correspond precisely to experiment. Nonetheless, thermodynamics must be considered when selecting candidates for frontal polymerization. Another example of thermodynamics limiting conversion is a ceiling temperature effect that can be seen in acrylamide polymerization fronts. A ΔT of 400°C can be seen when the T_i is that of liquid N_2, but when the T_i is 25 °C, a ΔT of only 260 °C is seen. (See Figure 7) If the front was allowed to reach its maximum ΔT, a higher conversion would be achieved.

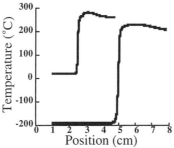

Figure 7. Spatial temperature profiles for acrylamide fronts with different initial temperatures. (Adapted from Pojman et al.) (12)

2. Molecular Weight Distributions Enikolopyan et al. analytically calculated the molecular weight distribution, MWD, assuming the reaction takes place at the front temperature and that the initiator is consumed (22). They predicted degrees of polymerization less than 150, with a maximum mass fraction at 30.

Gel permeation chromatography (GPC) showed that poly(methacrylic acid) from fronts had multimodal, broad MWDs with surprisingly high molecular weights (7). Given that the front temperatures exceed 200 °C and that the initiator concentrations are large (1%), it was unexpected that high molecular weight polymer was produced (> 10^5). Pojman et al. found in a recent study that these high molecular weights were the result of crosslinking via intermolecular anhydride formation (23). By analyzing the samples after anhydride cleavage, they found that the molecular weight dropped significantly (from $M_n = 1.4$ x 10^5 to $M_n = 1.0$ x 10^4).

Because acrylate esters cannot undergo an analogous reaction, the molecular weight distribution for poly(n-butyl acrylate) was determined (23). The MWDs followed the trend that one would expect from the classical steady-state theory of free-radical polymerization. Poly(n-butyl acrylate) produced in fronts had broad MWDs. ($M_w/M_n = 1.7 - 2.0$) The average molecular weight decreased with increasing initiator concentrations (23).

CAB-O-SIL was added to thermoplastic systems to avoid the Rayleigh-Taylor instability. It was not known what effect this inert filler would have on the MWD of the product. To determine this, Pojman et al produced poly(n-butyl acrylate) frontally in microgravity (in the absence of Rayleigh-Taylor instabilities), and the resulting MWD was compared with a product produced on Earth using CAB-O-SIL to suppress the convective instabilities (24). There was very little difference in the two MWDs.

III. Solid Monomers

Although the bulk of research in frontal polymerization has been performed in liquid systems (methacrylic acid, triethylene glycol dimethacrylate), there has also been research done with solid monomers. As previously stated, because of its relatively low melting point (in comparison to the front temperature) and its high reactivity, acrylamide is a good example of a solid monomer.

Industrial polymerization of acrylamide is done in solvent, usually water. Attempts have been made to produce polyacrylamide frontally. Our research group has demonstrated frontal polymerization with acrylamide and various initiators: benzoyl peroxide, lauroyl peroxide, azobisisobutyronitrile (AIBN). The current system being studied is acrylamide and 4% (w/w) potassium persulfate initiator.

Previous research with the acrylamide-potassium persulfate system includes studies done on the effect of particle size of monomer on front velocity, as well as the effect of green density (density of the unreacted pellet) on front velocity and front temperature (14,25).

The acrylamide-potassium persulfate system was polymerized frontally with no solvent to produce polyacrylamide by a solvent-free process. However, the high temperatures associated with frontal polymerization (230-260 °C) resulted in intermolecular imidization (26), which crosslinked the polymer chains and formed an insoluble polymer (Equation 9). The polymer formed was insoluble in water, a solvent that dissolves polyacrylamide. In order to avoid the intermolecular imidization that crosslinks the polymer, the front temperature was lowered by diluting the reaction mixture.

The diluents used were the product of the reaction, an inert filler (barium carbonate), and commercial polyacrylamide. The experimental procedure was as follows. In dilution with product, the product formed from frontal polymerization with acrylamide and potassium persulfate (4% w/w) was ground into a powder and added to the reaction mixture. The acrylamide, initiator, and product were mixed for thirty minutes. In dilution with inert filler, barium carbonate was added to acrylamide and potassium persulfate (4% w/w) and mixed for thirty minutes. In dilution with commercial polyacrylamide, commercial polyacrylamide was added to acrylamide and potassium persulfate (4% w/w) and mixed for thirty minutes. All dilutions were done in a ratio of 0.8:1 diluent to monomer. The samples were pressed with a hydraulic press to a pellet density of 1.1 g/cm^3. The pellets were placed in 16x150 mm test tubes, and the front was ignited with a soldering iron.

$$\text{⁖⁖⁖ CH}_2 - \text{CH ⁖⁖⁖}$$
$$|$$
$$\text{O} = \text{C}$$
$$|$$
$$\text{NH}_2$$
$$|$$
$$\text{NH}_2$$
$$|$$
$$\text{C} = \text{O}$$
$$|$$
$$\text{⁖⁖⁖ CH}_2 - \text{CH ⁖⁖⁖}$$

$\longrightarrow$

$$\text{⁖⁖⁖ CH}_2 - \text{CH ⁖⁖⁖}$$
$$|$$
$$\text{C} = \text{O}$$
$$|$$
$$\text{N H} \qquad +\text{NH}_3$$
$$|$$
$$\text{C} = \text{O}$$
$$|$$
$$\text{⁖⁖⁖ CH}_2 - \text{CH ⁖⁖⁖}$$

(9)

   To make velocity measurements, centimeter labels from Wale Apparatus were placed on 16 x 150 mm test tubes. A stopwatch was used to time the front. The value of the slope of the line represented the velocity in centimeters per second.
   To obtain temperature profiles, a hole one centimeter in depth was drilled into the top of the pellet with a 0.125 mm drill bit. Thermocouples of 0.125 mm in diameter from Omega were placed into the hole. The thermocouples and an on/off switch were connected to the digital input of an A/D board. Temperature and time were measured using Workbench software with the A/D board. The temperature vs. time profiles were converted to temperature vs. distance profiles using the velocities of the fronts.
   The undiluted acrylamide-potassium persulfate sample had a maximum temperature (T_{max}) of 235 °C as seen in Figure 8. All of the diluted samples had lower front temperatures than the undiluted sample.

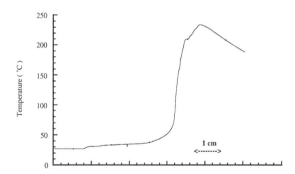

Figure 8. Temperature profile of an undiluted sample of acrylamide and potassium persulfate (4% w/w). The maximum temperature (T_{max}) is 235 °C.

   In the sample diluted with the product of the reaction, the maximum temperature (T_{max}) decreased from 235 °C in the undiluted sample to 165 °C. The diluent acts as a heat sink, absorbing heat away from the front. The same results were seen when the diluents were barium carbonate and commercial polyacrylamide (Figure 9). In the front diluted with barium carbonate, the maximum temperature (T_{max}) was as low as 95 °C.

148

Thermal gravimetric analysis (TGA) of the product, the filler-diluted sample, product-diluted sample, and the polyacrylamide-diluted sample showed evidence of intermolecular imidization in the sample that had been diluted with commercial polyacrylamide (Figure 10). There was a weight decrease of about 20% in the polyacrylamide-diluted sample that was not seen in the other samples at *ca.* 250 °C. This could be from intermolecular imidization releasing NH_3. Calculations show that if every amide in a sample of polyacrylamide is imidized, the resulting weight loss would be 17%. Imidization could be occurring in the commercial polyacrylamide diluent.

The polyacrylamide-diluted sample is water-soluble, as is polyacrylamide. However, further analytical tests must be done to confirm the structure of the product.

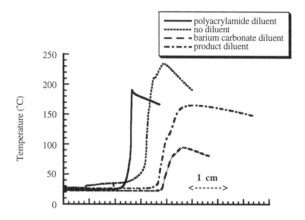

Figure 9. Temperature profile of acrylamide-potassium persulfate system (4% w/w) diluted with product (T_{max}=165 °C), inert filler (barium carbonate) (T_{max}=94 °C), and commercial polyacrylamide (T_{max}=190 °C). The T_{max} of the undiluted front was 235 °C.

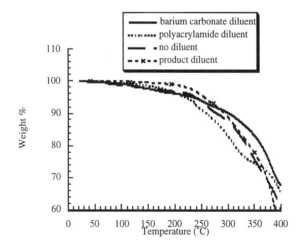

Figure 10. TGA of various samples showing weight loss in the polyacrylamide-diluted sample at *ca.* 250 °C because of imidization of commercial polyacrylamide.

IV. Interferences

As promising as the frontal polymerization approach is, there are two major interferences with its commercial application. Convection caused by the large density gradients created in a front can quench fronts. Even without convection, periodic modes of propagation, so-called spin modes, can leave spiral patterns in the product.

1. Convection Changes in density in polymerization reactions occur for two reasons. One reason that has already been established is from the exothermicity of the reaction (*27*). The other factor resulting in a change in density is a composition change (*27*). As the wavefront passes it is converting monomer into polymer, which is on the order of 20% more dense than the monomer. In the instance of a descending front this can have drastic effects and in some cases destroys the front. As the polymer is formed it is more dense than the monomer below it. Because of this it starts to sink, thereby distorting the front and in some cases destroying it. (See Figure 3.) This is referred to as the Rayleigh-Taylor instability (*28,29*). Currently the only method to prevent this instability is to add a filler.

Ascending fronts are plagued by convection caused by the large temperature gradient. Bowden et al. studied the critical conditions that lead to convection in ascending fronts of crosslinked polymer (*30*). Most importantly, they determined it was possible to eliminate convection by increasing the initial viscosity or increasing the front velocity. Garbey et al. found a similar result in theoretical analysis for ascending fronts with a liquid product (*31*).

Horizontally-propagating fronts also exhibit convection. Chekanov et al. found that for the frontal curing of epoxy, horizontal fronts were completely quenched by convection (*8*).

The diameter of the tube also affects the frontal process. The larger the tube, the smaller the density gradient necessary to induce convection and the faster the convective flow if it occurs. Because of this, narrowing the tube serves to stabilize the front and decrease the amount of convection. However, too narrow a tube can result in front extinction from heat loss.

2. Non-Uniform Propagation Not all systems are susceptible to convective instabilities, yet more generic instabilities exist that may arise in any thermally propagating front. For example, frontal polymerization of methacrylic acid or the frontal cationic curing of epoxies, lowering the initial temperature or increasing the amount of heat loss can cause the onset of periodic "spinning" modes of propagation (*12,32*). Figure 11 shows the remarkable spiral pattern spontaneously generated in the frontal polymerization of diacrylate at room temperature.

Figure 11. Images of a 1,6-Hexanediol diacrylate spin mode with Lupersol 231 initiator. An indicator, Bromophenol Blue, is used to accentuate the visibility of the spin mode. (Images courtesy of J. Masere)

If lowering the initial temperature was the only way to achieve spin modes, then it would be an interesting phenomenon to study, but would not hinder the use of frontal polymerization for material synthesis. However, this is not the case. For multifunctional acrylates, these "spinning" modes are observed even at room temperature if an initiator with a sufficiently high activation energy is used.

Maintaining the initial system above the critical temperature for the onset of the instability and reducing heat losses can prevent such instabilities, although it is usually not possible to know the critical temperature a priori. We do know three factors that favor periodic modes: high front temperature, high energy of activation and low initial temperature. These factors are summarized by the Zeldovich number (*16*):

$$Z = \frac{T_m - T_o}{T_m} \frac{E_{eff}}{RT_m} \tag{10}$$

For adiabatic systems with one step reaction kinetics, for all values of Z less than 8.4, planar fronts occur, but Solovyov et al. have found that models with realistic polymerization kinetics are more stable than predicted by Equation 9 (*33*).

V. Basic Aspects of Isothermal Frontal Polymerization

Frontal polymerization is a mode of converting monomer into polymer via a localized propagating reaction zone. Isothermal Frontal Polymerization (IFP), also known as Interfacial Gel Polymerization, occurs when a polymer seed, i.e. a small piece of polymer, is placed in contact with a solution including its monomer and a thermal initiator. When the solution containing monomer and initiator come in contact with the polymer seed, the monomer and initiator diffuse into the seed. This diffused region allows polymerization to occur faster due to the Tromsdorff effect (*34*) resulting in a propagating front.

The primary industrial use of isothermal fronts is in the production of gradient optical materials such as Gradient Refractive INdex (GRIN) materials. A GRIN material is one in which a gradient in the refractive index exists over the thickness of the material. GRINs have many uses, such as protection of equipment and people from high intensity laser beams, magnifying materials, optical fiber amplifiers, and optical lenses (*35,36*). One example of a GRIN material which exhibits magnifying properties is shown in Figure 12. There are two methods for developing GRINs in a polymer medium, the first involves the use of copolymers and the second involves the use of a dopant material. The gradients in both methods arise from differences in diffusion rates of the two materials involved: The copolymer gradient is produced from different reactivity rates of the two polymers, where one monomer polymerizes more slowly than the other, thus, diffusing into the polymer matrix more slowly. The dopant gradient is produced from the dopant diffusing into the polymer seed more slowly than the monomer. Koike et al. have developed procedures for producing optical lenses, or disks, using copolymer techniques and also have developed production procedures for fiber amplifiers using the dopant technique (*35,36*), while Zhang et al. have produced polymer preforms using the dopant technique (*37*).

While GRIN materials are in use, there is not a clear understanding of the factors that affect the process of IFP (*38-41*). Therefore, we have undertaken this project. While IFP has not been studied as extensively as Thermal Frontal Polymerization (TFP), there are several trends that do appear including the time and distance for the front to propagate. These two trends will be mentioned, as well as several properties of the system, how these properties affect the front, and several preliminary experimental results.

1. Experimental Process IFP occurs when a polymer seed is placed in contact with a solution of its monomer and a thermal initiator. The polymer seed is made in a test tube by isothermally heating (e.g. with a water bath) an amount of monomer until it homogeneously polymerizes. The seed remains in the isothermal system for a period of approximately thirty-six hours to ensure high percent conversion and high molecular weight. A bulk solution, containing monomer, a thermal initiator, and sometimes an inhibitor, is added to the test tube and placed in a thermostated bath.

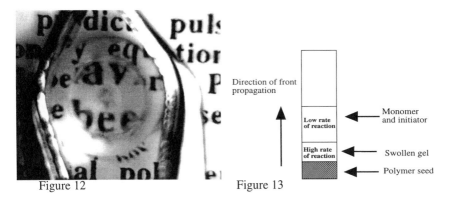

Figure 12   Figure 13

Figure 12. Magnification by GRIN material composed of poly(methyl methacrylate) with naphthalene dopant.

Figure 13. IFP in swollen gel

The monomer and initiator diffuse into the top portion of the polymer seed causing it to swell. The viscosity in this swelled portion of seed is much higher than in bulk solution: The Tromsdorff effect (the gel effect or auto-acceleration) is the process where the high viscosity of the seed slows termination reactions in the polymerization of free-radicals while not significantly affecting the rate of chain growth (34). Because the swelled gel has a slower rate of chain termination, the monomer that diffuses into the seed continues to grow, not homogeneously polymerizing as does its counterpart in the bulk solution. Because of this continuation in growth, a propagating front is produced.

When an inhibitor is used in the bulk solution, the rate of propagation is enhanced. The inhibitor does not diffuse into the polymer matrix as quickly as the monomer and initiator, especially if the inhibitor is of a larger size than the polymer matrix. The inhibitor prohibits, or slows, the formation of the polymer in the bulk solution creating a bigger difference between the rates of polymerization in the seed and in the bulk solution allowing the front to propagate further.

The question of whether this propagation is actually frontal polymerization or if it is only due to swelling of the seed, can be answered with experimental data. The seed can swell from the monomer, but only a certain amount. Any fronts that travel a length exceeding the amount that the seed can swell, will prove conclusively that the system is undergoing IFP. Work is underway in our lab to provide this proof.

2. Trends and Properties of IFP Even though there is not a clear understanding of the mechanism behind IFP, there is sufficient work to suggest several trends for this process and also to hypothesize how several properties will affect the system. Current work in our lab shows that the system is highly dependent upon the combination of monomer, initiator, and inhibitor. Systems containing methyl methacrylate (MMA) with various combinations of initiators, including lauryl peroxide (LPO) and azobisisobutylnitrile (AIBN), and inhibitors, including poly-4-vinylphenol (P4VP) and TEMPO (2,2,6,6-tetramethyl-1-piperidinyloxy), have been in use. The specific combination, and also the concentrations, determine the experimental temperature of the system based on the reactivities of the monomer, initiator, and inhibitor. A higher temperature is desired because it results in a faster rate of reaction and, thus, a faster propagation, as well as a higher molecular weight of the polymer.

Further results from our lab using the previously mentioned systems of MMA with LPO and MMA with AIBN, both with P4VP or TEMPO show that the average distance of propagation is 1.0+/-0.3 cm with the average time of travel approximately twenty-four to thirty-six hours depending on the system. Using these systems, we

152

have not been able to accurately gauge the linearity of the velocity, but we suspect that it is non-linear as first order initiator rate of consumption calculations predict.
Besides the dependence of the system on temperature and combination of monomer, initiator, and inhibitor, it is suspected that convection plays an important role in IFP. As in TFP, density gradients occur in the monomer solution. Buoyancy-driven convection should mix the bulk solution resulting in a homogeneous density. Even so, isothermal fronts can only be ascending fronts, unlike their thermal counterparts, because the polymer is more dense than the monomer and would collapse if in a descending front.

VI. Conclusions

Thermal frontal polymerization is a rapid method for producing polymeric materials without solvent and without the danger associated with bulk polymerization of such reactive materials as acrylates. The front velocity can be controlled by the type of initiator and its concentration. Relatively low molecular weight polymers are produced. Not all monomers can be used because of interfering convective instabilities caused by the large compositional and thermal gradients produced in a front.
Isothermal frontal polymerization is a slow method to produce high molecular polymer using a gradient of reaction rate in a polymeric gel in contact with its monomer. Not enough is known about the factors that affect front propagation but the method is used to produce gradient materials.

References

1. Chechilo, N. M.; Enikolopyan, N. S. *Dokl. Phys. Chem.* **1975**, *221*, 392-394.
2. Chechilo, N. M.; Enikolopyan, N. S. *Dokl. Phys. Chem.* **1976**, *230*, 840-843.
3. Pojman, J. A. *J. Am. Chem. Soc.* **1991**, *113*, 6284-6286.
4. Chechilo, N. M.; Khvilivitskii, R. J.; Enikolopyan, N. S. *Dokl. Akad. Nauk SSSR* **1972**, *204*, 1180-1181.
5. Chechilo, N. M.; Enikolopyan, N. S. *Dokl. Phys. Chem.* **1974**, *214*, 174-176.
6. Pojman, J. A.; Khan, A., M.; West, W. *Polym. Prepr. Am Chem. Soc. Div. Polym. Chem.* **1992**, *33*, 1188-1189.
7. Pojman, J. A.; Willis, J.; Fortenberry, D.; Ilyashenko, V.; Khan, A. *J. Polym. Sci. Part A: Polym Chem.* **1995**, *33*, 643-652.
8. Chekanov, Y.; Arrington, D.; Brust, G.; Pojman, J. A. *J. Appl. Polym. Sci.* **1997**, *66*, 1209-1216.
9. Manelis, G. B.; Smirnov, L. P. *Combust. Explos. Shock Waves* **1976**, *5*, 665.
10. Pojman, J. A.; Craven, R.; Khan, A.; West, W. *J. Phys. Chem.* **1992**, *96*, 7466-7472.
11. Nagy, I. P.; Pojman, J. A. *J. Phys. Chem.* **1996**, *100*, 3299-3304.
12. Pojman, J. A.; Ilyashenko, V. M.; Khan, A. M. *J. Chem. Soc. Faraday Trans.* **1996**, *92*, 2825-2837.
13. Pojman, J. A.; Nagy, I. P.; Salter, C. *J. Am. Chem. Soc.* **1993**, *115*, 11044-11045.
14. Fortenberry, D. I.; Khan, A.; Ilyashenko, V. M.; Pojman, J. A. in *Synthesis and Characterization of Advanced Materials, ACS Symposium Series No. 681*; Serio, M. A.; Gruen, D. M. Malhotra, R.,Ed., American Chemical Society: Washington, DC, 1998; pp 220-235.
15. Khan, A. M.; Pojman, J. A. *Trends Polym. Sci. (Cambridge, U.K.)* **1996**, *4*, 253-257.
16. Zeldovich, Y. B.; Barenblatt, G. I.; Librovich, V. B.; Makhviladze, G. M. *The Mathematical Theory of Combustion and Explosions*; Consultants Bureau: New York, 1985.

17. Goldfeder, P. M.; Volpert, V. A.; Ilyashenko, V. M.; Khan, A. M.; Pojman, J. A.; Solovyov, S. E. *J. Phys. Chem. B* **1997**, *101*, 3474-3482.
18. Pojman, J. A.; Elcan, W.; Khan, A. M.; Mathias, L. *J. Polym. Sci. Part A: Polym Chem.* **1997**, *35*, 227-230.
19. Davtyan, S. P.; Zhirkov, P. V.; Vol'fson, S. A. *Russ. Chem. Rev.* **1984**, *53*, 150-163.
20. Sawada, H. *Thermodynamics of Polymerization*; Marcel Dekker: New York, 1976.
21. Odian, G. *Principles of Polymerization*; Wiley: New York, 1991.
22. Enikolopyan, N. S.; Kozhushner, M. A.; Khanukaev, B. B. *Doklady Phys. Chem.* **1974**, *217*, 676-678.
23. Pojman, J. A.; Willis, J. R.; Khan , A. M.; West, W. W. *J. Polym. Sci. Part A: Polym Chem.* **1996**, *34*, 991-995.
24. Pojman, J. A.; Khan, A. M.; Mathias, L. *J. Microg. sci. tech. in press.*
25. Pojman, J.; Fortenberry, D.; Ilyashenko, V. *Int. J. SHS* **1997**, *6*, 355-376.
26. Sandler, S. R.; Karo, W. *Polymer Syntheses*; Academic Press: New York, 1974.
27. Pojman, J. A.; Epstein, I. R. *J. Phys. Chem.* **1990**, *94*, 4966-4972.
28. Rayleigh, J. W. *Scientific Papers, ii*; Cambridge University Press: Cambridge, 1899.
29. Taylor, G. *Proc. Roy. Soc. (London)* **1950**, *Ser. A. 202*, 192-196.
30. Bowden, G.; Garbey, M.; Ilyashenko, V. M.; Pojman, J. A.; Solovyov, S.; Taik, A.; Volpert, V. *J. Phys. Chem. B* **1997**, *101*, 678-686.
31. Garbey, M.; Taik, A.; Volpert, V. *Quart. Appl. Math.* **1996**, *54*, 225-247.
32. Pojman, J. A.; Ilyashenko, V. M.; Khan, A. M. *Physica D* **1995**, *84*, 260-268.
33. Solovyov, S. E.; Ilyashenko, V. M.; Pojman, J. A. *Chaos* **1997**, *7*, 331-340.
34. Trommsdorff, E.; Köhle, H.; Lagally, P. *Makromol. Chem.* **1948**, *1*, 169-198.
35. Yasuhiro Koike, E. N. US Patent 5,253,323, 1993.
36. Keisuke Sasaki, Y. K. US Patent 5,450,232, 1995.
37. Q. Zhang, P., Zhang, Yan Zhai *Macromolecules* **1997**, *30*, 7874-7879.
38. V. V. Ivanov, E. V. S., and L. M. Pushchaeva *Chem. Phys. Reports* **1997**, *16*, 947-951.
39. Smirnov, B. R. ;. M., S. S.; Lusinov, I. A.; Sidorenko, A. A.;Stegno, E. V.; Ivanov, V. V. *Vysokomol. Soedin., Ser. B*, **1993**, *35*, 161-162.
40. Golubev, V. B.; Gromov, D. G.; Korolev, B. A. *J. Appl. Polymer Science* **1992**, *46*, 1501-1502.
41. Gromov, D. G.; Frisch, H. L. *J. Appl. Polymer Science* **1992**, *46*, 1499-1500.

SYNTHESIS AND PROCESSING
IN SUPERCRITICAL CARBON DIOXIDE

Chapter 10

Chain Growth Polymerizations in Liquid and Supercritical Carbon Dioxide

Murat A. Quadir[1] and Joseph M. DeSimone[2,3]

Department of Chemistry, Venable Hall, CB 3290, University of North Carolina at Chapel Hill, Chapel Hill, NC 27599–3290

Abstract

The use of carbon dioxide as an alternative to traditional solvents has emerged recently as an important development in polymer chemistry. The past year has witnessed major advances in the synthesis of a variety of polymeric materials in carbon dioxide, with primary emphasis on utilizing carbon dioxide as an inert reaction medium. The scope of the studies conducted in our laboratories on chain-growth polymerizations covers free radical, cationic, and transition-metal-catalyzed polymerizations in liquid and supercritical carbon dioxide. These systems have included homogeneous and heterogeneous (including precipitation and dispersion) polymerizations, as well as hybrid carbon dioxide/aqueous phases. As will be shown, an array of polymers can be readily prepared using these polymerization methods in the environmentally friendly medium of liquid and supercritical carbon dioxide.

[1]mquadir@email.unc.edu.
[2]desimone@unc.edu.
[3]Corresponding author.

Introduction

Solution polymerizations are generally conducted in organic solvents or in aqueous media. Large amounts of volatile organic compounds (VOCs), chlorofluorocarbons (CFCs), and hazardous aqueous waste streams are used as reaction media or during manufacturing, and are released as a result of industrial productions. Many emulsion and suspension polymers are often spray-dried or oven-dried after synthesis which literally involves trillions of BTUs every year. Recently, reducing or eliminating the emission of toxic organic compounds and polluted aqueous waste streams has become a global issue. Since it has intrinsic environmental advantages, carbon dioxide is useful as a benign solvent alternative. It is nontoxic, nonflammable, and can be easily separated and recycled. In addition, CO_2 is inexpensive and widely available, allowing the potential for large-scale industrial use. The utilization of solvents such as CO_2 would provide an excellent opportunity for a more environmentally responsible chemical industry.

The tunability of solvency of carbon dioxide by simply altering pressure and temperature makes it a very attractive reaction medium. In most cases, CO_2 only acts as a good solvent at relatively high pressure in the liquid or supercritical state.[1] A supercritical fluid (SCF) can be defined as a substance that is above its critical temperature (T_c) and critical pressure (P_c). Under these conditions the substance does not behave as a typical gas or liquid but exhibits hybrid properties of these two states. Near the supercritical point, small perturbations in temperature and pressure result in large changes in density, viscosity, and dielectric constant. It is therefore possible to tune the density of a SCF isothermally, simply by raising or lowering the pressure (Figure 1). Since changes in pressure cause CO_2 to vary from vapor-like to liquid-like, one can control solvent properties over a considerable range. From a practical standpoint, CO_2 has rather modest supercritical parameters $(T_c = 31.1\ °C, P_c = 73.8\ bar)$ and supercritical conditions are therefore easily obtained. Supercritical carbon dioxide $(scCO_2)$ resembles gas phase behavior such as high mixing rates, miscibility with other gases, and relatively weak molecular association, while maintaining liquid properties, especially the ability to dissolve and transport organic compounds.[2] For these reasons carbon dioxide is a viable alternative to traditional solvents. Furthermore, polymers synthesized in such a medium do not need to undergo costly drying or solvent removal procedures. It should be noted that the use of $scCO_2$ as the monomer that lies beyond the scope of this paper. Additionally, there is a large amount of literature which has shown that CO_2 can be used as a comonomer for various polymerizations, particularly with epoxides to form aliphatic polycarbonates. As yet, there are only preliminary reports of the use of $scCO_2$ as both monomer and solvent in a copolymerization process[3, 4], although this is likely to be an exciting area for future research since CO_2 is a very inexpensive C_1 feedstock.

In contrast to ordinary organic solvents, the solvent power of $scCO_2$ is different for polymers than for small molecules. There is an empirical rule that states that if a low molecular weight compound dissolves in hexane then it is

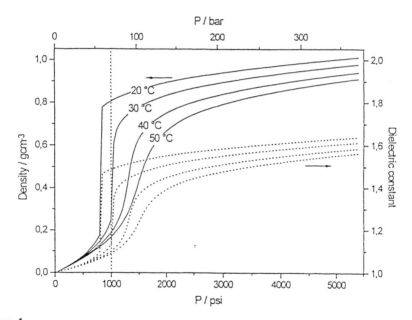

Figure 1.
Density (solid line) and dielectric constant (dotted line) of CO_2 as a function of pressure and temperature. The vertical line represents the critical pressure of CO_2 at 1030 psia (71 bar).

soluble in carbon dioxide. However, for high molecular weight substances such as polymers, the dissolving power of carbon dioxide is more like that of a fluorocarbon. Thus many industrially important hydrocarbon polymers are insoluble in carbon dioxide and must be polymerized by heterogeneous reactions. By contrast, fluorocarbon and silicone polymers are soluble in carbon dioxide and can be synthesized via homogeneous methods.

Carbon dioxide was first shown to be a good solvent for free radical precipitation polymerizations in 1968.[5] The free radical precipitation polymerizations of vinyl chloride, styrene, methyl methacrylate, acrylonitrile, acrylic acid, and vinyl acetate were carried out in liquid and scCO$_2$ and resulted in variable yields. These reactions were initiated with either γ radiation or common free radical initiators. Since the polymers which were formed are insoluble in CO$_2$, these reactions were precipitation reactions and the products were recovered as powders or gums. A number of free radical precipitation polymerizations in scCO$_2$ have been discussed in the literature. Polyethylene was prepared at ca. 400 bar and temperatures between 20 and 45 °C with high molecular weight (10^5 g/mol).[6] In 1986, acrylic acid and vinyl C$_3$ to C$_5$ carboxylic acids were polymerized to form high molecular weight materials.[7] A copolymerzation of vinyl carboxylic acids and divinyl monomers resulted in crosslinked poly(acrylic acids).[8] There are even reports that alpha-olefins were polymerized in CO$_2$ despite the fact that they do not polymerize well by free radical processes under ordinary conditions because of the degradative chain transfer of allylic hydrogen from the monomer to the propagating radical.[9] All of these reports indicate that CO$_2$ is effectively inert to free radical polymerization which avoids the issue of chain transfer to solvent.

Another intrinsic advantage of using carbon dioxide as a polymerization solvent is the fact that CO$_2$ can plasticize and swell many polymeric materials,[1, 10-12] allowing one to tune not only the solvent environment but potentially the glass transition temperature (T_g), melting temperature, and viscosity of the polymer-rich particles being formed. This plasticization affects the growing polymer particle and allows diffusion of monomer into the particle interior[13-17] and facilitating the incorporation of additives into the polymer.[18-21] CO$_2$ significantly swells amorphous polymers such as polystyrene, thereby suppressing the T_g. Wang, Kramer, and Sachse studied the effects of high pressure CO$_2$ on the glass transition temperature and mechanical properties of polystyrene by measuring Young's modulus and creep compliance.[22] Chiuo, Barlow, and Paul estimated the T_g of PMMA, PS, PC, PVC, PET, and blends of PMMA and PVF$_2$ plasticized by CO$_2$ with pressures up to 25 bar by differential scanning calorimetry (DSC) and observed reductions of T_g by up to 50 °C.[23] In addition to plasticizing the polymer and facilitating diffusion of reactants, CO$_2$ diffuses rapidly from the polymer matrix upon de-pressurization thus leaving no solvent residues.

There are other advantages in the processing of polymers with CO$_2$. One is the opportunity to extract residual monomer, solvent, or catalyst from a solid polymer.[1] Second, a mixture of polymers with different molecular weights can be fractionated by change the density of supercritical carbon dioxide.[24-27] Finally, the polymer morphology can be controlled with supercritical drying[28] or foaming.[29-31]

At the University of North Carolina at Chapel Hill we mainly explore the feasibility of CO_2 as an inert medium for polymerization. Since 1992, we have investigated free radical, cationic, and transition-metal-catalyzed polymerizations in liquid and $scCO_2$. These polymerizations have included homogeneous and heterogeneous (precipitation and dispersion) systems, as well as hybrid carbon dioxide/aqueous phases. At the same time, complementary studies (high pressure NMR, small angle X-ray and neutron scattering techniques, and dynamic light scattering) have helped to successfully elucidate the physical behavior of a range of polymers in carbon dioxide. A clear understanding of the physical chemistry of CO_2-based polymerization systems enables us to design more effective surfactants and stabilizers for use in CO_2, and to predict their behavior in $scCO_2$ solution. The synthetic methodology developed in our group has demonstrated that environmentally friendly liquid and supercritical carbon dioxide can be applied as an effective medium for the synthesis of a large number of polymers (Table 1).[32-51] This paper provides primarily an overview of our current research work, while research emphasis on synthesis in liquid and supercritical carbon dioxide in the period 1992-1996 have been reviewed in great detail elsewhere.[52, 53]

Homogeneous Solution Polymerizations in Carbon Dioxide

Free Radical Chain Propagation
The high solubility of fluorocarbons in carbon dioxide makes CO_2 an attractive alternative solvent for the synthesis and processing of fluorinated polymers. DeSimone *et al.* reported the first homogeneous polymerization using CO_2 for partially fluorinated acrylate monomers,[54] such as 1,1-dihydroperfluorooctylacrylate (FOA) (Scheme 1). A number of fluorinated polymers as well as some hydrocarbon/fluorocarbon copolymers were also synthesized homogeneously, including poly[2-(*N*-methyl-perfluoro-octanesulfon-amido)ethyl acrylate], poly[2-(*N*-ethyl-perfluoro-octanesulfonamido)ethyl acrylate], poly[2-(*N*-ethyl-perfluorooctanesulfonamido)ethyl methacrylate], poly(1,1-dihydroperfluorooctyl methacrylate), poly(*p*-perfluoroethyleneoxy methylstyrene), poly(FOA-*co*-styrene), poly(FOA-*co*-methyl methacrylate), poly(FOA-*co*-butyl acrylate), and poly(FOA-*co*-ethylene),[54, 55] as indicated in Table 2. Homogeneous copolymerization of FOA with 2-(dimethylamino) ethylacrylate or with 4-vinylpyridine has also been employed to prepare CO_2-soluble polymeric amines.[50]

The thermal decomposition kinetics of free radical initiator AIBN has also been studied using high-pressure ultraviolet spectroscopy.[56] It was shown that the

Table 1.
Summary of selected polymerization reactions in liquid / supercritical carbon dioxide.

Monomer	Phase*	Pressure (bar)	Temp. (°C)	$<M_n>$ (kg/mol)	PDI[†]	Yield (%)	Ref.
$CF_2=CF_2$ (TFE)	H/P	165–350	68–180	0.6–900[‡]	1.35–1.44[§]	13–900	32–35
$CF_2=CFO(CF_2)CF_3$ (PPVE)/TFE	P	90–110	35	>1000[#]	n/a	99–100	36
$CF_2=CF(CF_3)$ (HFP)/TFE	P	50–100	35	>1000[¶]	n/a	3–82	36
$CH_2=CF_2$ (VF₂)	H	280–340	60	~0.6[¥]	1.05	32–35	37
$CH_2=CH-C(O)O-CH_2(CF_2)_6CF_3$ (FOA)	H	207	60	11–1600[**]	n/a	65	33, 38, 39
$CH_2=CH-Ph$ (styrene)	D/P[††]	204–345	65	3.8–84.3	1.4–5.3	12–98	40–42
$CH_2=CH(CH_3)C(O)O-CH_3$ (MMA)	D/P[††]	69–350	30–65	65–390	2.1–5.3	10–98	16, 33, 43, 44
$(CH_3)_2C=CH_2$ (isobutylene)	P	75–135	32.5–36	0.5–2.5	1.5–3.4	5–35	45
$CH_2=CH-CH-COOH$ (acrylic acid)	P	125–345	62	2.9–153	1.3–3.9	>90	46
$CH_2=CH-C(O)-NH_2$ (acrylamide)	E	340	60	4900–7000[‡‡]	n/a	91.4–99.8	33, 47
Norbornene	P	68–345	65	15–511	2.0–4.5	6–75[§§]	33, 48
Cyclic/vinyl ethers	H/P[##]	290–345	-10–60	7–152	1.2–9.4	43–91	33, 49
2,6-Dimethylphenylene oxide	P/D[¶¶]	345	25–40	0.9–17.2	1.6–5.8	16–86	50

*Phase behavior during polymerization: H = homogeneous, P = precipitation, D = dispersion, E = emulsion. [†]PDI: polydispersity index of molecular weight distribution. [‡]Reaction condition as both telomerization and polymerization. [§]PDI refers to telomerization reaction [32]. [#]Incorporation of PPVE in copolymer = 2.9-8.6 wt.%. [¶]Incorporation of HFP in copolymer = 11.2-13.8 wt.%. [¥]Telomerization reaction. [**]Molecular weights have been determined by both gel permeation chromatography (GPC) [43] and small angle neutron scattering (SANS) [51]. [††]A dispersion is only observed in the presence of a stabilizer. [‡‡]Viscosity average molecular weight (M_v) is reported. [§§]Methanol used as a cosolvent in some reactions. [##]Only fluorinated systems were found to be homogeneous. [¶¶]Some evidence for dispersion in presence of stabilizer, at least in early stages of reaction.

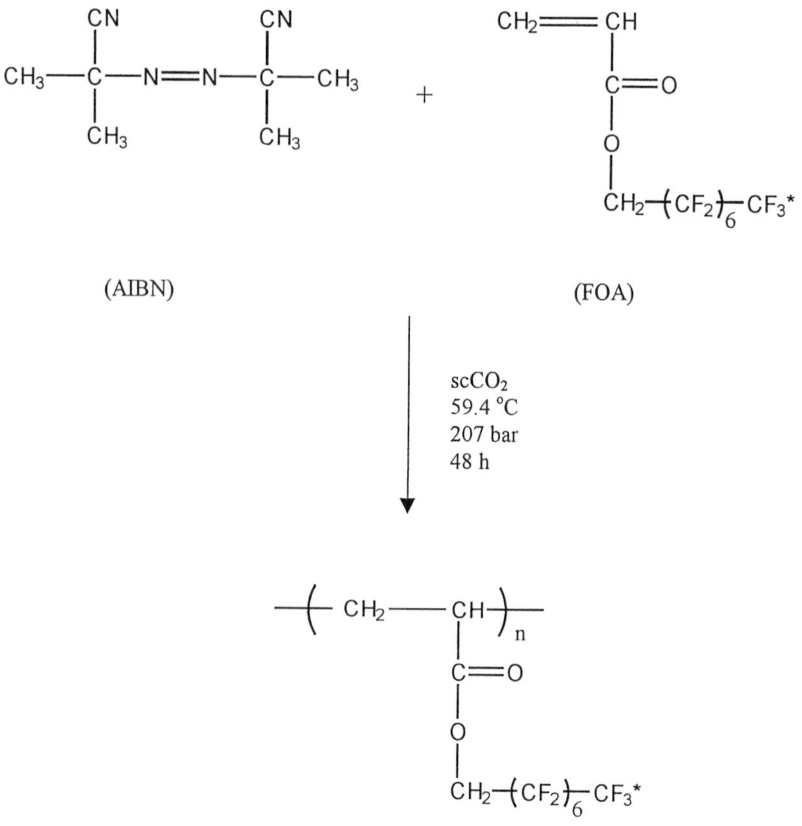

Scheme 1.
Homopolymerization of FOA in CO_2 (*Contains *ca.* 25% -CF_3 branches).[54]

Table 2.
Statistical copolymers of FOA with vinyl monomers. Polymerizations were conducted at $59.4 \pm 0.1\ ^{\circ}C$ and 345 ± 0.5 bar for 48 hours in CO_2. Intrinsic viscosities were determined in 1,2-trifluoroethane (Freon-113) at $30\ ^{\circ}C$.[54, 55]

Copolymer	Feed Ratio	Incorporated	Intrinsic Viscosity (dL/g)
poly(FOA-*co*-MMA)	0.47	0.57	0.10
poly(FOA-*co*-styrene)	0.48	0.58	0.15
poly(FOA-*co*-BA)	0.53	0.57	0.45
poly(FOA-*co*-ethylene)	0.35	-	0.14

decomposition kinetics and initiator efficiency in CO_2 differ from those measured in other organic media. The rate constant for AIBN decomposition was *ca.* 2.5 times slower in CO_2 than in benzene. This was attributed to the low dielectric constant of carbon dioxide relative to benzene. As expected, initiator efficiency was found to be very high (> 80%) as consequence of negligible solvent cage effects in supercritical CO_2 resulting from its low viscosity. This work concluded that CO_2 is indeed an inert medium to free radicals, and therefore an excellent solvent for conducting free radical polymerizations. Recently we have preformed the first successful pulsed laser polymerization (PLP) measurement of the free radical propagation rate coefficients (k_p) of styrene and methyl methacrylate in $scCO_2$[57]. PLP in combination with size exclusion chromatography (SEC) for the determination of k_p has been recommended by the IUPAC working party on *"Radical Polymerization Kinetics and Processes"*.[58-60] The k_p values measured at 65 °C with a pressure at *ca.* 200 bar show that the propagation rate coefficients of styrene and methyl methacrylate in supercritical carbon dioxide do not change as compared to the bulk propagation rate coefficients corrected for pressures with the volume of activation (Table 3). In principle, the solvent effect for k_p is not expected to be large. This indicates that bulk propagation rate coefficients apparently can be used to estimate propagation rates in carbon dioxide.

The inertness of CO_2 to free radicals is further exemplified by our work with the telomerization of tetrafluoroethylene (TFE).[32] Using AIBN as an initiator gave inconsistent results due to the high degree of chain transfer of the oligomeric TFE radicals to initiator or other hydrocarbons that were present. When perfluorobutyl iodide was used as the telogen and reactions were conducted in the absence of the thermal initiator AIBN (Scheme 2), better results were obtained. As Table 4 illustrates, the perfluorlkyl iodides prepared in this manner had both controlled molecular weights and lower polydispersity indexes (PDI). The separation of polymer from homogenous CO_2 solution was facile. Upon venting CO_2 from the reaction vessel, the polymer precipitated from solution leaving a dry, solvent-free product. In addition, dilution of highly reactive monomers such as TFE may be handled more safely when diluted as a mixture with CO_2.[32, 61]

Cationic Chain Propagation
Our group reported the first homogeneous cationic polymerizations in liquid and supercritical carbon dioxide.[49] Two vinyl ethers containing fluorinated side chains were polymerized at 40 °C in $scCO_2$ using ethylaluminum dichloride as an initiator (Lewis acid) (Scheme 3). The typical yield of these polymers was around 40%, and the polymer bearing the fluorinated sulfonamide side chain had a number average molecular weight ($< M_n >$) of 4.5 x 10^3 g/mol with a PDI of 1.6. Additionally, the ring-opening polymerization of a flurorinated oxetane, 3-methyl-3'-[(1,1-dihydroheptafluoro-butoxy)methyl]oxetane (FOx-7), was performed homogeneously in liquid CO_2 at 0 °C and 289 bar (Scheme 4) using trifluoroethanol as the initiator and boron trifluoride tetrahydrofuranate as the coinitiator. Similar results were found in experiments that were conducted in

Table 3.
Propagation rate coefficients in supercritical carbon dioxide obtained by the PLP method.[57]

Monomer	Pressure (bar)	Temperature (°C)	k_p^{obs} $M^{-1}s^{-1}$	k_p^{theor} $M^{-1}s^{-1}$
MMA	162	66.5	1150	1069
MMA	156	66.4	1020	1062
MMA	196	67.0	905	1103
Styrene	187	65.1	420	441
Styrene	164	65.8	467	447
Styrene	181	65.1	494	440

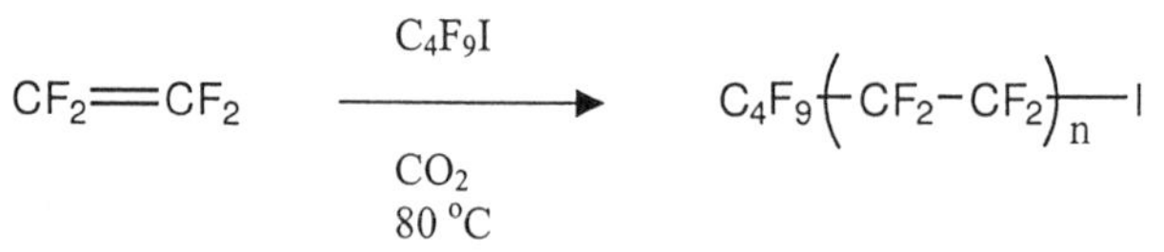

$$CF_2{=}CF_2 \quad \xrightarrow[\substack{CO_2 \\ 80\,°C}]{C_4F_9I} \quad C_4F_9{\left(CF_2{-}CF_2\right)}_n{-}I$$

Scheme 2.
Telomerization of TFE in supercritical CO_2.[32]

Table 4.
Experimental results from the telomerizations of TFE in supercritical CO_2. Perfluorobutyl iodide was employed as the telogen.[32]

[monomer]/[telogen]	Yield (%)	$< M_n >$ (g/mol)	PDI
1.6	88	570	1.35
1.5	87	590	1.38
1.8	86	630	1.38
2.2	78	650	1.44

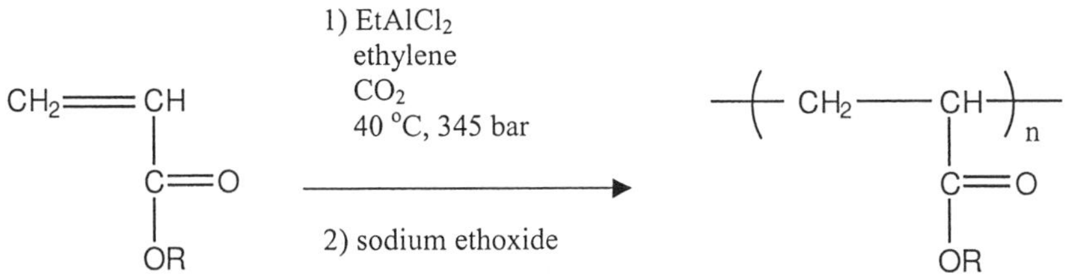

CH_2══CH

    |
    C══O
    |
    OR

1) $EtAlCl_2$
   ethylene
   CO_2
   40 °C, 345 bar

———————————→

2) sodium ethoxide

$-(-CH_2$——$CH-)_n-$

            |
            C══O
            |
            OR

$R = -CH_2CH_2(CF_2)_nCF_3;$ $n = 5 - 7$

$R = -CH_2CH_2N(C_3H_7)SO_2C_8F_{17}$

Scheme 3.
Homogeneous polymerization of fluorinated vinyl ethers in supercritical CO_2.[49]

1) BF_3-THF
   CO_2
   0 °C, 289 bar
2) NaOH

$H-(-O-CH_2$——$\overset{\displaystyle CH_3}{\underset{\displaystyle CH_2OCH_2CF_2CF_2CF_3}{C}}$——$CH_2-)_n-OH$

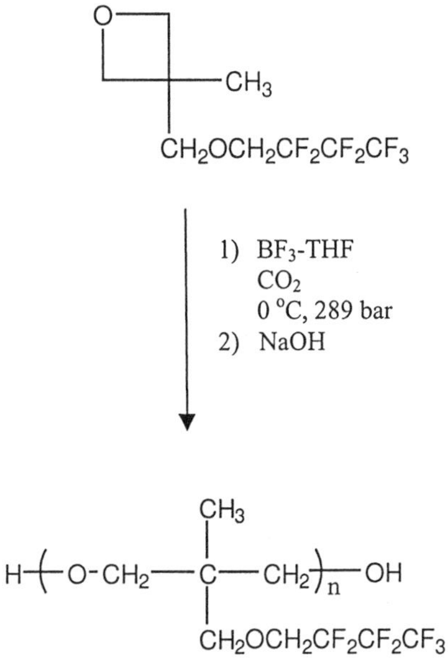

Scheme 4.
Homogeneous cationic polymerization of Fox-7 in supercritical CO_2.[49]

Freon-113, supporting the analogy of CO_2 to CFCs and suggesting that the former certainly will be an excellent alternative in solution applications.

Heterogeneous Polymerizations in Carbon Dioxide

Precipitation Polymerizations
We have studied the precipitation polymerization of acrylic acid in supercritical carbon dioxide, and have found that using ethyl mercaptan as a chain transfer agent effectively controls molecular weight.[46] Greater than 90% conversion was observed when AIBN was employed as an initiator. The polymers were recovered as a fluffy white powder. We have recently reported the copolymerizations of TFE with perfluoro(propyl vinyl ether) (PPVE) and also with hexafluoropropylene (HFP) in $scCO_2$.[36, 62] In both cases, bis(perfluoro-2-propoxy propionyl)peroxide was used as the free radical initiator. As shown in Table 1, good yields of high molar mass copolymers resulted from these polymerizations. The advantages of using this CO_2-based system are threefold. First, chain transfer does not occur between CO_2 and the highly electrophilic monomers derived from fluoroolefin compounds. Secondly, undesirable carboxylic acid and acid fluoride end group that are formed in the aqueous process, causing processing and performance problems, are not observed in CO_2. Finally, as mentioned before, safer handling of TFE can be accomplished as a mixture with CO_2.[61]

Dispersion Polymerizations
Typically, a free radical dispersion polymerization begins homogeneously, since monomer and initiator present are soluble in the continuous phase. When the growing polymer chain reaches a critical length j_{crit}, phase separation occurs.[63, 64] Lipophilic and hydrophilic polymers are less soluble in CO_2 and these polymers have been described as "CO_2-phobic". The polymerizations of acrylic acid and styrene, as well as the copolymerization of TFE in $scCO_2$, are good recent examples (Table 1). In the absence of a stabilizer or solubilizing agent, CO_2-phobic polymers will precipitate rapidly from CO_2 solution as they are formed. In a dispersion polymerization, an interfacial agent such as a stabilizer or a surfactant can stabilize the resulting polymer as a colloid. This stabilization provides higher degrees of polymerization than the corresponding precipitation polymerization and forms spherical particles with sizes usually in the range of microns up to several millimeters, facilitating workup and processing. Since vinyl olefins are miscible in CO_2 whereas their polymers are insoluble, they can be polymerized via dispersion polymerizations.

Prevention of coagulation or flocculation of colloidal dispersion particles can occur by use of electrostatic, and steric stabilization.[65-67] Since CO_2 is a low dielectric medium, steric stabilization by polymer chains is the preferred method for dispersion polymerizations in CO_2. The most effective steric stabilizers are amphiphilic molecules that become adsorbed or grafted onto the surface of the polymer particle through an anchoring segment. The molecules also contain stabilizing segments that are soluble in the continuous phase. The polymeric stabilizer is a macromolecule that preferentially exists at the polymer-solvent interface and prevents aggregation of particles by coating the surface of the particles and imparting long-range repulsions between them. These long range interactions must be large enough to compensate for long range *van der Walls* attractions.[66] For a stabilizer to be effective, a proper balance must be found between the sections that are soluble and insoluble in the continuous phase.

We have completed the first successful dispersion polymerization of a lipophilic monomer in $scCO_2$ by polymerizing methyl methacrylate using steric stabilization.[43, 68, 69] The amphiphilic nature of poly(FOA) (Figure 2) was used as a stabilizer for the synthesis of poly(methyl methacrylate) (PMMA). The lipophilic, acrylic-like chain anchors the growing polymer particle while the perfluorinated segment extends into the continuous phase.[43] The results from the AIBN initiated polymerizations are shown in Table 5. The PMMA produced in this manner was in the form of nearly monodispersed particles with sizes ranging from 0.9 to 2.7 μm, as characterized by scanning electron microscopy (SEM). Very low amounts (0.24 wt%) of poly(FOA) are needed to stabilize PMMA as latexes in CO_2,[16, 70] and the expensive stabilizer can often be reclaimed with a CO_2 wash. The ability of CO_2 to plasticize PMMA facilitates diffusion of monomer into the growing polymer particles which allows the reaction to proceed to high conversion. Conversion greater than 90% and a molecular weight of about 3×10^5 g/mol was observed for these PMMA particles. Polymerization performed in the absence of poly(FOA) resulted in low conversion and low molar masses.

Another effective dispersion polymerization of MMA involved the copolymerization of MMA with a polydimethylsiloxane (PDMS) (Figure 3) monomethacrylate macromonomer, which acted as the stabilizer.[44] The polymerizations were performed in both liquid and $scCO_2$ with AIBN initiator and resulted in spherical particles. Only a small amount of PDMS was incorporated into the polymer to stabilize the particle, and excess PDMS monomer was removed from the product by washing with hexanes or CO_2.

We also extended dispersion polymerization to polystyrene,[41] which is an important polymer in plastic and elastomer applications. Poly(FOA) was found not to be an effective stabilizer for the dispersion polymerization of styrene, probably because of ineffective anchoring of the stabilizer in the growing polymer particle. Block copolymers consisting of a polystyrene anchoring segment, and a poly(FOA) block, which is readily solubilized in CO_2, were synthesized and used as effective stabilizers for the polymerization of styrene (Figure 4a).[71] These stabilizers were synthesized by the 'iniferter" method developed by Otsu.[72] Styrene was polymerized radically using tetraethylthiuram disulfide (TD) as the iniferter. To ensure high functionality, the polymerization was stopped at about 20% conversion. The block copolymers were generated by photochemical polymerization of FOA using TD-polystyrene as the macroiniferter by irradiation

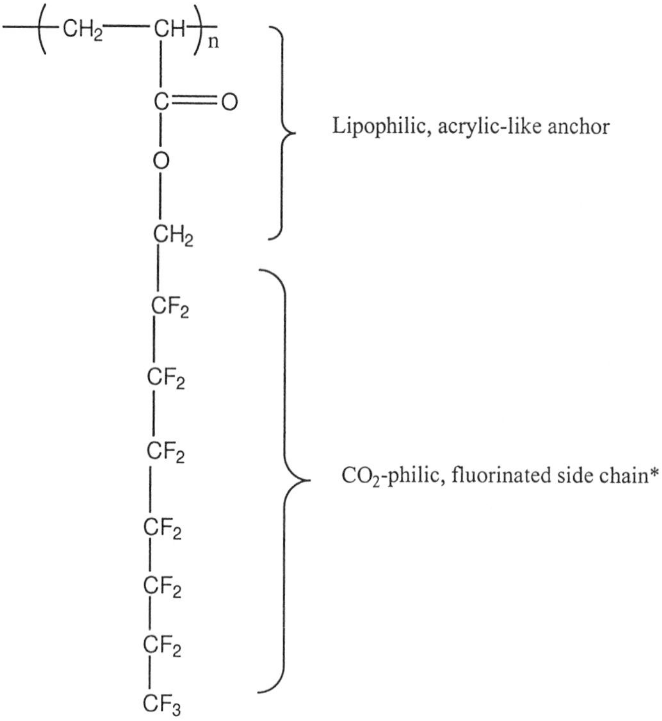

Figure 2.
Structure of amphiphilic polymeric stabilizer poly(FOA) (*Contains *ca.* 25% -CF$_3$ branches).[43]

Table 5.
Results of MMA polymerization with AIBN as the initiator in CO$_2$ at 204 bar and 65 °C; stabilizer is either low molecular weight (LMW) or high molecular weight (HWM) poly(FOA).[43]

Stabilizer (w/v %)	Yield (%)	$<M_n>$ (10^{-3} g/mol)	PDI	Particle Size (μm)
0%	39	149	2.8	-
2% LMH	85	308	2.3	1.2 (± 0.3)
4% LMH	92	220	2.6	1.3 (± 0.4)
2% HMW	92	315	2.1	2.7 (± 0.1)
4% HMW	95	321	2.2	2.5 (± 0.2)

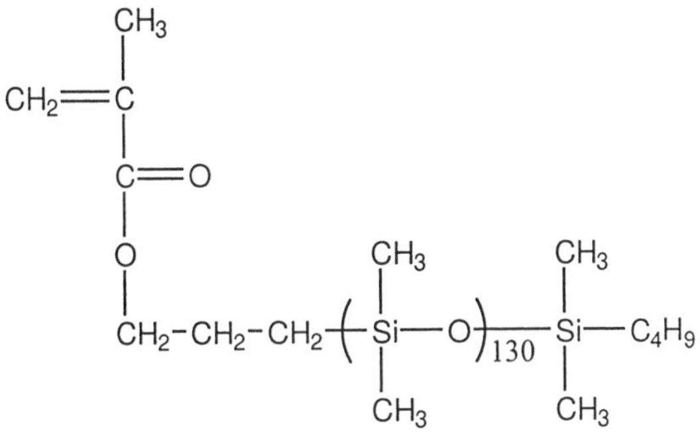

Figure 3.
Structure of PDMS macromonomer used a copolymerizable stabilizer.[44]

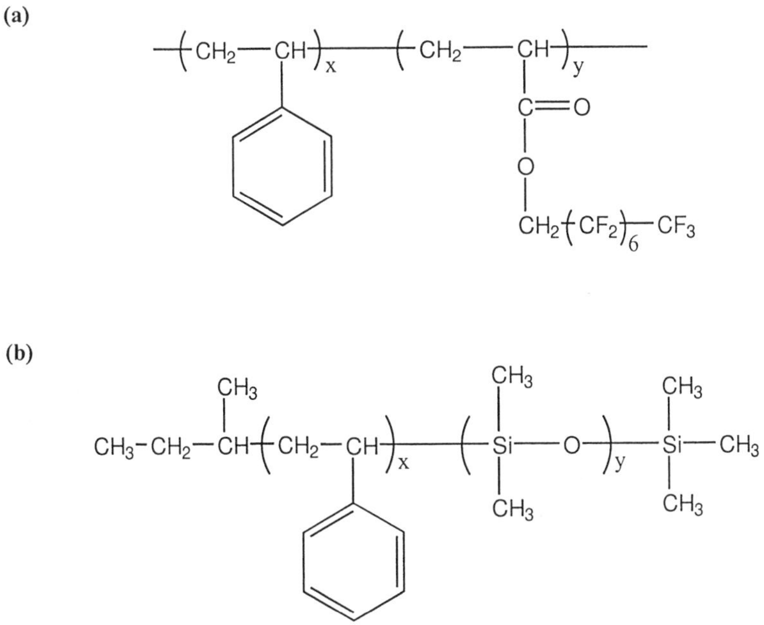

(a)

(b)

Figure 4.
Structures of amphiphilic diblock copolymeric stabilizers: (a) PS-*b*-PFOA, (b) PS-*b*-PDMS.[68]

with a mercury lamp. Using this PS/Poly(FOA) copolymer as a stabilizer for the polymerization of styrene resulted in dispersion instead of precipitation polymerization. The results from these polymerizations are summarized in Table 6. The polymer was recovered as dry powder and was in the form of submicron sized spherical particles as demonstrated by SEM. The size of these particles was dictated by variation in the PS segments and the poly(FOA) segments in the stabilizer.

Block copolymers of PS and PDMS (polydimethylsiloxane) were also effectively employed as stabilizers in the dispersion polymerization of styrene (Figure 4b).[42] An advantage of this system is the much lower cost of the PDMS as compared to PFOA. It was also discovered that PS-b-PDMS was an effective stabilizer in the polymerization of MMA. The resulting PMMA particles were about an order of magnitude smaller than the particles formed using PFOA as the stabilizer. PDMS homopolymer was ineffective in stabilizing these polymerization reactions.

The cationic dispersion polymerization of styrene in liquid carbon dioxide using amphiphilic block copolymers was also investigated in our group.[73] Block copolymers bearing a CO_2-philic poly(fluorinated vinyl ether) segment and a poly(methyl vinyl ether) segment stabilized the growing polystyrene particles. SEM pictures showed the formation of PS particles in the size range of 300 nm to 1 μm in diameter.

Ring-Opening Metathesis in Carbon Dioxide

Heterogeneous ring-opening metathesis polymerizations have been carried out in carbon dioxide and CO_2 cosolvent systems.[48, 74] In this work, bicyclo[2.2.1]hept-2-ene (norbornene) was polymerized in CO_2 and CO_2/methanol mixtures using a $Ru(H_2O)_6(tos)_2$ initiator (Scheme 5). These reactions were conducted at 65 °C with pressures ranging from 60 to 345 bar; resulting in poly(norbornene) with similar conversions and molecular weights as those obtained in other solvent systems. ^{1}H-NMR spectroscopy showed that the product prepared in pure CO_2 had the same structure as the product obtained from pure methanol (Figure 5). More interestingly, it was shown that the cis/trans ratio of the polymer microstructure could be controlled by the addition of a methanol cosolvent to the polymerization medium. The poly(norbornene) made in pure methanol or in methanol/CO_2 mixtures had a very high trans-vinylene content, while the polymer made in pure CO_2 had a very high cis-vinylene content. These results can be explained by the solvent effects on relative populations of the two different possible metal carbene propagating species, which give rise to the different microstructures. Investigations involving the use of other organometallic initiating systems and other cyclic olefin monomers are underway. Indeed, the expansion of polymer synthesis in CO_2 to include metal catalyzed reaction mechanisms represents an important step in the use of CO_2 as a universal solvent for polymerization.

Table 6.
Results of styrene polymerization with AIBN as the initiator in CO_2 at 204 bar and 65 °C; stabilizer is either poly(FOA) homopolymer or PS-*b*-PFOA copolymer.[41]

Stabilizer	Yield (%)	$<M_n>$ $(10^{-3}$ g/mol)	PDI	Particle Diameter (μm)	Particle Size Distribution
none	22.1	3.8	2.3	none	none
poly(FOA)	43.5	12.8	2.8	none	none
3.7K/16K	72.1	19.2	3.6	0.40	8.3
4.5K/25K	97.7	22.5	3.1	0.24	1.3
6.6K/35K	93.6	23.4	3.0	0.24	1.1

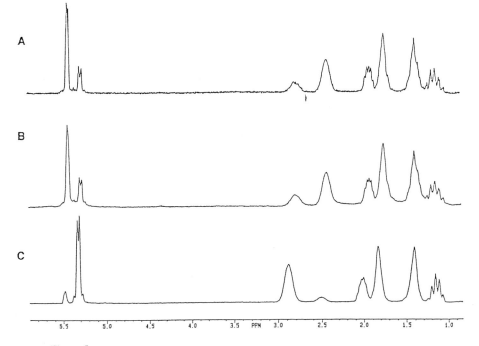

Scheme 5.
Ring-opening methathesis polymerization of norbornene in CO_2.[48, 73]

Figure 5.
^{1}H-NMR of poly(norbornene) prepared in: **A)** methanol, **B)** methanol/CO_2, and **C)** CO_2.[73]

Hybrid CO_2/Aqueous Medium

Because carbon dioxide and water exhibit low mutual solubilities, the combination of a CO_2-based process and a conventional aqueous heterogeneous polymerization process is similar to the biphasic fluorocarbon/hydrocarbon system for olefin hydroformylation recently reported by Horvath.[75] This system provides a substantial departure from current technologies, allowing for the compartmentalization of monomer, polymer, and initiator based on their solubility characteristics. Furthermore, the tunability inherent to $scCO_2$ allows variation of the density, viscosity, and solvent strength of the carbon dioxide phase with simple changes in pressure or temperature. It is well known that carbon dioxide significantly plasticizes many polymer particles,[1, 10-12] allowing one to tune not only the solvent environment but potentially the T_g, melting temperature, and viscosity of the polymer-rich particles being formed. This could produce qualitative changes in kinetics, molecular weight distributions, etc., since it could lead to different mechanisms for reactions such as termination which are normally diffusion-controlled.[64] This might result in qualitatively different molecular weight distributions (MWD) with changing pressure of CO_2, with a concomitant reduction in the gel effect being manifest in the kinetics. The advantages of using CO_2, rather than a more common liquid organic solvent are threefold. First, environmental considerations make CO_2 a much more responsible choice for a reaction medium than liquid organic solvents. Secondly, solvent removal is achieved simply by careful venting of the system. Lastly, CO_2 has a viscosity that is, under the conditions used, significantly lower than that of conventional liquid solvents, and would thus be expected to have a greater impact on reaction kinetics. These biphasic systems are ideally suited for handling highly reactive monomers such as chloroprene or TFE as detailed in the previous section, where monomer dilution with inert carbon dioxide can serve to moderate reactivity. Also, the high heat capacity of water can be advantageous in controlling exothermic polymerizations. In this newly developed system, TFE was polymerized by using a water-soluble initiator at 75 °C and sodium perfluorooctanoate was used as the surfactant. From this process, reasonably high molar mass ($< M_n > = 1$ x 10^6 g/mol) PTFE was produced in good yield (80-90%).[34, 76-78]

Recently, we have reported the detailed characterization of a surfactant-free methyl methacrylate emulsion polymerization using such a hybrid system.[17] The reaction mechanism for a traditional surfactant-free emulsion polymerization in an aqueous continuous phase with an initiator such as persulfate is generally referred to as following a "homogeneous-coagulation nucleation" process.[17, 64, 79, 80] An initiator molecule in the aqueous phase undergoes aqueous-phase propagation and termination, until it reaches a degree of polymerization z when it can either enter a pre-existing particle or grow further to a degree of polymerization j_{crit}. An oligomeric radical of size j_{crit} may homogeneously nucleate to form a new (precursor) particle. Precursor particles are colloidally unstable, and grow by both propagation and by coagulation until they achieve sufficient size and charge density to become stable to further coagulation.

Colloidal stability is provided by *in situ* surfactant, i.e., the endgroups arising from initiator dissociation (which are either oligomeric species formed from aqueous-phase termination, or grafted long chains). Because the amount of such *in situ* surfactant is comparatively small, latex particles formed in a "surfactant-free" emulsion polymerization are usually large. Interval 1 denotes the period of particle formation, when newly formed radicals may form new particles. Interval 2 denotes the period of particle growth in the absence of new particle formation (i.e., when the only fates available to newly formed radicals are either aqueous-phase termination or entry into a pre-existing particle) in the presence of monomer droplets. Interval 3 denotes the final period of particle growth, when droplets have disappeared and monomer is contained only in the particle and (usually to a much small extent) in the continuous phase.

A hybrid system comprises water and high-pressure CO_2 is depicted in Scheme 6. Visual inspection of the system showed that there were always two continuous phases present (in addition to the particle phase), a water-rich and a CO_2-rich phase. These phases remained separate, and under low agitation a clear phase boundary was seen between the upper CO_2-rich phase and the lower water-rich phase. Polymer particles were observed only in the water-rich phase, with no visible creaming (accumulation at an interface). In this work, surfactant-free polymerizations of methyl methacrylate have been carried out at 75 °C using potassium persulfate under different CO_2 pressures, yielding poly(methyl methacrylate) as a stable latex. Figure 6 shows a typical SEM of PMMA from the surfactant-free polymerization conducted in the absence of CO_2 and in the presence of CO_2. It is apparent that the latex particles are spherical and free from agglomeration. More striking is the observation that the particle size is smaller in the presence of CO_2 than without supercritical carbon dioxide present. The particle size distribution has also been measured by capillary hydrodynamic fractionation (CHDF), and the particle number concentrations, N_c, have then been obtained. An interesting result that emerges from the data on the particle numbers is that there is no systematic trend in N_c with CO_2 pressure or with initiator concentration (Figure 7). This observation has also been reported in surfactant-free MMA systems without CO_2.[81]

Conversion–time data lend credibility to the hypothesis that the gel effect is reduced when the reaction is carried out with significant amounts of CO_2. There is also a general, but not systematic, tendency to lower rate at high pressures; the slower rate could thus be explained by a lower radical population caused by decreased viscosity within particles at higher pressures. Typical GPC distributions are given in Figure 8.

In the future, the use of CO_2 in conjunction with a second immiscible solvent phase could become very important. The addition of $scCO_2$ to a conventional surfactant-free emulsion polymerization has a definite effect upon both the reaction kinetics and the product formed. In particular, the sound and profound change in the molecular weight distribution at higher CO_2 pressure is due to a fundamental changes in chain-stopping mechanism. Indeed, this is an intrinsic advantage in the heterogeneous polymerizations mentioned above: CO_2 swells the growing polymer particle and allows diffusion of monomer into the

174

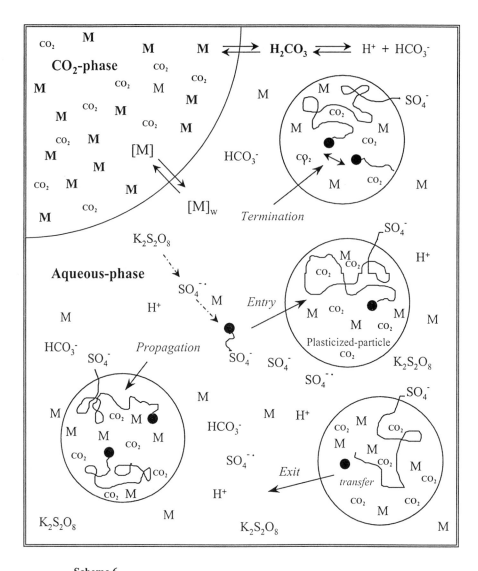

Scheme 6.

Schematic of emulsion polymerization in a hybrid CO_2 / aqueous medium.[17]

a

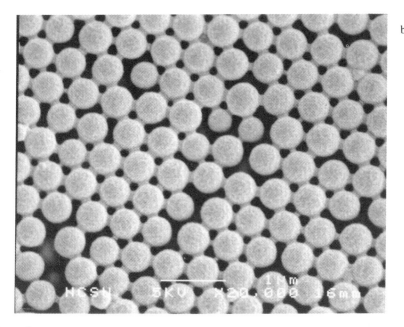

b

Figure 6.
Scanning electron micrographs for particles formed: a) without CO_2, (b) with CO_2 at 113 bar. Both samples are taken at *ca.* 80% conversion.[17]

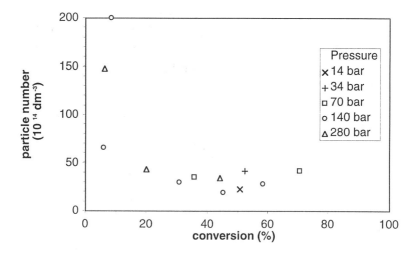

Figure 7.

Dependence of particle number on conversion for MMA hybrid CO_2/water system initiated by 1 mM potassium persulfate at various pressures of CO_2 as indicated. High values of particle number at low conversion are probably artifacts.[17]

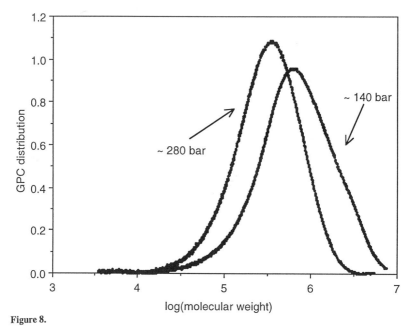

Figure 8.

GPC distributions (which are GPC traces with baseline subtracted and de convoluted to remove effects of a nonlinear calibration curve), for two samples from hybrid CO_2/water emulsion polymerization of MMA at indicated CO_2 pressures, conversion 45%. [17]

particle interior. Other explorations along this line target forming novel composite materials, in which a second monomer is polymerized inside a CO_2-swollen polymer host.[13-15] In addition to plasticizing the polymer and facilitating diffusion of reactants, CO_2 diffuses rapidly from the polymer matrix upon depressurization, leaving no solvent residues.[87-89]

Conclusions

Recent progress in our group as well as others have shown that CO_2 is an attractive alternative to conventional organic solvents in the manufacture and processing of polymers. Since CO_2 is inert to free radicals, it is particularly advantageous in free radical reactions. We have also shown CO_2 to be a good medium for cationic and ring-opening metathesis polymerizations. Stabilizers designed for CO_2 extend the possibility of polymerization to a wide array of monomers via dispersion polymerization. The ability to tailor the surfactant molecule for a particular CO_2-based application may prove to be the enabling technology for reduction in the use of hazardous VOCs and CFCs and for the elimination of the use of water to save trillions of BTUs from drying processes. The synthesis of polymer blends in CO_2-swollen substrates can also be achieved. An interesting extension of this area would be to exploit the plasticization effects of $scCO_2$ in order to synthesize composite polymer materials with more rigidly defined structures, molecular weights and molecular weight distributions. The effectiveness of CO_2 as polymerization medium, coupled with its environmental responsibility, makes it a solvent of choice for future manufacturing of polymers.

Acknowledgements

We gratefully acknowledge the National Science Foundation (J. M. DeSimone, Presidential Faculty Fellow, 1993-1998), and the Kenan Center for the Utilization of Carbon Dioxide in Manufacturing, sponsored by Rohm and Haas, B. F. Goodrich, Becton Dickinson, Oak Ridge National Laboratory, Eastman Chemical Company, DuPont, Solvay, Atochem, Japan Synthetic Rubber Co. LTD, MiCELL Technologies, Air Products and Chemicals, Oxychem, Praxair and Phasex.

References

(1)   McHugh, M. A.; Krukonis, V. J. *Supercritical Fluid Extraction*; (2nd Ed.) Butterworth-Heinemann, **1994**.

(2)   Jessop, P. G.; Ikariya, T.; Noyori, R. *Nature* **1994**, *368*, 231.

(3)   Costello, C. A.; Berluche, E.; Han, S. J.; Sysyn, D. A.; Super, M. S.; Beckman, E. J. *Am. Chem. Soc. Polym. Mater. Sci. Eng.* **1996**, *74*, 430

(4)   Super, M. S.; Beckman, E. J. *Trends. Polym. Sci.* **1997**, *5*, 236.
(5)   Sumitomo French Patent 1,524,533, **1968**.
(6)   Hagiwara, M.; Mitsui, H.; Machi, S.; Kagiya, T. *J. Polym. Sci.*: Part A-1, **1968**, *6*, 603.
(7)   Sertage, W. G. Jr.; Davis, P.; Schenck, H. U.; Denzinger, W.; Hartman, H. Canadian Patent 1,274,942, **1986**.
(8)   Hartmann, H.; Denzinger, W. Canadian Patent 1,262,995, **1986**.
(9)   Terry, R. E.; Zaid, A.; Angelos, C.; Whitman, D. L. *Energy Prog.* **1988**, *8*, 48.
(10)   Condo, P. D.; Johston, K. P. *J. Polym. Sci. Part B Polym. Phys.* **1994**, *32*, 523.
(11)   Briscoe, B. J.; Kelly, C. T. *Polymer* **1995**, *36*, 3099.
(12)   Kwag, C.; Gerhardt, L. J.; Khan, V.; Gulari, E.; Manke, C. W. *Polym. Mater. Sci. Eng.* **1996**, *74*, 183.
(13)   Watkins, J. J.; McCarthy, T. J. *Macromolecules* **1994**, *27*, 4845.
(14)   Watkins, J. J.; McCarthy, T. J. *Macromolecules* **1995**, *28*, 4067.
(15)   Watkins, J. J.; McCarthy, T. J. *Polym. Mater. Sci. Eng.* **1996**, *74*, 402.
(16)   Hsiao, Y. L.; Maury, E. E.; DeSimone, J. M.; Mawson, S. M.; Johnston, K. P. *Macromolecules* **1995**, *28*, 8159.
(17)   Quadir, M. A.; Snook, R.; Gilbert, R. G.; DeSimone, J. M. *Macromolecules* **1997**, *30*, 6015.
(18)   Sand, M. L. United States Patent 4,598,006, **1986**.
(19)   Berens, A. R.; Huvard, G. S.; Korsmeyer, R. W. United States Patent 4,820,752, **1989**.
(20)   Berens, A. R.; Huvard, G. S.; Korsmeyer, R. W.; Kunig, F. W. *J. Appl. Polym. Sci.* **1992**, *46*, 231.
(21)   Watkins, J. J.; McCarthy, T. J. *Polym. Mater. Sci. Eng.* **1995**, *73*, 158.
(22)   Wang, W. C.; Kramer, E. J.; Sachse, W. H. *J. Polym. Sci.: Polym. Phys. Ed.* **1982**, *20*, 1371.
(23)   Chiou, J. S.; Barlow, J. W.; Paul, D. R. *J. Appl. Polym. Sci.* **1985**, *30*, 2633.
(24)   Krukonis, V. *Polymer News* **1985**, *11*, 7.
(25)   McHugh, M. A.; Krukonis, V. J.; Pratt, J. A. *Trends Polym. Sci.* **1994**, *2*, 301.
(26)   Elsbernd, C. S.; DeSimone, J. M.; Hellstern, A. M.; Smith, S. D.; Gallagher, P. M.; Krukonis, V. J.; McGrath, J. E. *Polym. Prepr. (Am. Chem. Soc., Div. Polym. Chem.)* **1990**, *31*, 673.
(27)   Zhao, X.; Watkins, R.; Barton, S. W. *J. Appl. Polym. Sci.* **1995**, *55*, 773.
(28)   Elliot, J.; Jarrell, R.; Srinivasan, G.; Dhanuka, M.; Akhaury, R. United States Patent 5,128,382, **1992**.
(29)   Srinivasan, G.; Elliot, J. J. *Ind. Eng. Chem. Res.* **1992**, *31*, 1414.
(30)   Wessling, M.; Borneman, Z.; Van Den Boomgaard, T.; Smolders, C. A. *J. Appl. Polym. Sci.* **1994**, *53*, 1497.
(31)   Goel, S. K.; Beckman, E. J. *AIChE Journal* **1995**, *41*, 357.
(32)   Romack, T. J.; Combes, J. R.; DeSimone, J. M. *Macromolecules* **1995**, *28*, 1724.
(33)   Shaffer, K. A.; DeSimone, J. M. *Trends Polym. Sci.* **1995**, *3*, 146.
(34)   Romack, T. J.; Kipp, B. E.; DeSimone, J. M. *Macromolecules* **1995**, *28*, 8432.

(35) Kipp, B. E.; Romack, T. J.; DeSimone, J. M. *Polym. Mater. Sci. Eng.* **1996**, *74*, 264.
(36) Romack, T. J.; DeSimone, T. M.; Treat, T. A. *Macromolecules* **1995**, *28*, 8429.
(37) Combes, J. R.; Guan, Z.; DeSimone, J. M. *Macromolecules* **1994**, *27*, 865.
(38) DeSimone, J. M.; Guan, Z.; Elsbernd, C. S. *Science* **1992**, *257*, 945.
(39) Maury, E. E.; Batten, H. J.; Killian, S. K.; Menceloglu, Y. Z.; Combes, J. R.; DeSimone, J. M. *ACS Polym. Prepr.* **1993**, *34*, 664.
(40) O'Dell, P. G.; Hamer, G. K. *Polym. Mater. Sci. Eng.* **1996**, *74*, 404.
(41) Canelas, D. A.; Betts, D. E.; DeSimone, J. M. *Macromolecules* **1996**, *29*, 2818.
(42) Canelas, D. A.; Betts, D. E.; DeSimone, J. M. *Polym. Mater. Sci. Eng.* **1996**, *74*, 400.
(43) DeSimone, J. M.; Maury, E. E.; Menceloglu, Y. Z.; McClain, J. B.; Romack, T. J.; Combes, J. R. *Science* **1994**, *265*, 356.
(44) Shaffer, K. A.; Jones, T. A.; Canelas, D. A.; DeSimone, J. M. *Macromolecules* **1996**, *29*, 2704.
(45) Prenecker, T.; Kennedy, J. P. *Polym. Bull.* **1994**, *32*, 537.
(46) Romack, T. J.; Maury, E. E.; DeSimone, J. M. *Macromolecules* **1995**, *28*, 912.
(47) Adamsky, F. A.; Beckman, E. J. *Macromolecules* **1994**, *27*, 312.
(48) Mistele, C. D.; Thorp, H. H.; DeSimone, J. M. *J. Macromol. Sci. – Pure Appl. Chem.* **1996**, A33, 953.
(49) Clark, M. R.; DeSimone, J. M. *Macromolecules* **1995**, *28*, 3002.
(50) Kapellen, K. K.; Mistele, C. D.; DeSimone, J. M. *Macromolecules* **1996**, *29*, 495.
(51) McClain, J. B.; Londono, D.; Combes, J. R.; Romack, T. J.; Canelas, D. A.; Betts, D. E.; Wignall, G. D.; Samulski, E. T.; DeSimone, J. M. *J. Am. Chem. Soc.* **1996**, *118*, 917.
(52) Shaffer, K. A.; DeSimone, J. M. *Trends Polym. Sci.* **1995**, *3*, 146.
(53) Canelas, D. A.; DeSimone, J. M. *Adv. Polym. Sci.* **1996**, *133*, 142
(54) DeSimone, J. M.; Guan, Z.; Elsbernd, C. S. *Science* **1992**, *257*, 945.
(55) Guan, Z.; Elsbernd, C. S.; DeSimone, J. M. *Polym. Prepr. (Am. Chem. Soc. Div. Polym. Chem.)* **1992**, *34*, 329.
(56) Guan, Z.; Combes, J. R.; Menceloglu, Y. Z.; DeSimone, J. M. *Macromolecules* **1993**, *26*, 2663.
(57) van Herk, A. M.; Manders, B. G.; Canelas, D. A.; Quadir, M. A.; DeSimone, J. M. *Macromolecules* **1997**, *30*, 4780.
(58) Gilbert, R. G. *Pure Appl. Chem.* **1996**, *68*, 1491.
(59) Buback, M.; Gilbert, R. G.; Hutchinson, R. A.; Klumperman, B.; Kuchta, F. D.; Manders, B. G.; O'Driscoll, K. F.; Russell, G. T.; Schweer, J. *Macromol. Chem. Phys.* **1995**, *96*, 3267.
(60) Beuermann, S.; Buback, M.; Davis, T. P.; Gilbert, R. G.; Hutchinson, R. A.; Olaj, O. F.; Russell, G. T. Schweer, J.; van Herk, A. M. *Macromol. Chem. Phys.* **1997**, *198*, 1545.
(61) van Bramer, D.J.; Shiflett, M. B.; Yokozeki, A. United States Patent 5,345,013, **1994**.
(62) Romack, T. J.; DeSimone, J. M. *Polym. Mater. Sci. Eng.* **1996**, *74*, 428.

180

(63) Arshady, R. *Colloid Polym. Sci.* **1992**, *270*, 717.
(64) Gilbert, R. G. *Emulsion Polymerization: A Mechanistic Approach*; Academic Press, London, **1995**.
(65) Napper, D. H. *Polymeric Stabilization of Colloidal Dispersions*; Academic Press, New York, 1983.
(66) Barrett, K. E. J. *Dispersion Polymerization in Organic Media*; John Wiley, New York, **1975**.
(67) Piirma, I. *Polymeric Surfactants*; Dekker, New York, **1992**.
(68) DeSimone, J. M.; Maury, E. E.; Combes, J. R.; Menceloglu, Y. Z. United States Patent 5,382,623, **1995**.
(69) DeSimone, J. M.; Maury, E. E.; Combes, J. R.; Menceloglu, Y. Z. United States Patent 5,312,882, **1995**.
(70) Hsiao, Y. L.; Maury, E. E.; DeSimone, J. M. *Polym. Prepr. (Am. Chem. Soc. Div. Polym. Chem.)* **1995**, *36*, 190.
(71) Guan, Z. B.; DeSimone, J. B. *Macromolecules* **1994**, *27*, 5527.
(72) Otsu, T.; Yoshida, M. *Makromol. Chem. Rapid Commun.* **1982**, *3*, 127.
(73) Clark, M. R.; Kendall, J. L.; DeSimone, J. M. *Macromolecules* **1995**, *30*, 6011.
(74) Mistele, C. D.; Thorp, H. H.; DeSimone, J. M. *Polym. Prepr (Am. Chem. Soc. Div. Polym. Chem.)* *36*, 507.
(75) Horvath, T. J.; Rabai, J. *Science* **1994**, *266*, 72.
(76) DeSimone, J. M.; Romack, T. J. *Multi-phase polymerization Process. 1.*, U.S. Patent 5,514,759 5,514,759, **1996**.
(77) DeSimone, J. M.; Romack, T. J. *Multi-phase polymerization Process. 2.*, U.S. Patent 5,527,865 5,527,865, **1996**.
(78) DeSimone, J. M.; Romack, T. J. *Multi-phase polymerization Process. 3.*, U.S. Patent 5,530,077 5,530,077, **1996**.
(79) Fitch, R. M.; Tsai, C. H. *Polymer Colloids*; Fitch, R. M., Ed.; Plenum: New York, **1971**, pp 73.
(80) Ugelstad, J.; Hansen, F. K. *Rubber Chem. Technol.* **1976**, *49*, 536.
(81) Tanrisever, T.; Okay, O.; Sonmezoglu, I. C. *J. Appl. Poly. Sci.* **1996**, *61*, 485.
(82) Clay, P. A.; Gilbert, R. G. *Macromolecules* **1995**, *28*, 552.
(83) Shortt, D. W. *J. Liquid Chromat.* **1993**, *16*, 3371.
(84) Russell, G. T.; Gilbert, R. G.; Napper, D. H. *Macromolecules* **1992**, *25*, 2459.
(85) Ballard, M. J.; Napper, D. H.; Gilbert, R. G. *J. Polym. Sci. Polym. Chem. Ed.* **1984**, *22*, 3225.
(86) Clay, P. A.; Gilbert, R. G.; Russell, G. T. *Macromolecules* **1997**, *30*, 1935.
(87) Howdle, S. M.; Ramsay, J. M.; Cooper, A. I. *J. Polym. Sci. Part B Polym. Phys.* **1994**, *32*, 541.
(88) Shieh, Y. T.; Su, J. H.; Manivannan, G.; Lee, P. H. C.; Sawan, S. P.; Spall, W. D. *J. Appl. Polym. Sci.* **1996**, *59*, 695.
(89) Shieh, Y. T.; Su, J. H.; Manivannan, G.; Lee, P. H. C.; Sawan, S. P.; Spall, W. D. *J. Appl. Polym. Sci.* **1996**, *59*, 707.

Chapter 11

Generation of Microcellular Biodegradable Polymers Using Supercritical Carbon Dioxide

D. Sparacio and E. J. Beckman

Chemical Engineering Department, University of Pittsburgh, Pittsburgh, PA 15261

In this study, a pressure quench method employing supercritical carbon dioxide was utilized to produce microcellular poly(lactic-co-glycolic acid) (PLGA). Saturation temperatures and pressures were varied to determine effects on average cell size, foam bulk density, and cell number density. PLGA foams were produced having average cell sizes ranging from 10 to 70 μm and cell number densities from 10^6 to 10^9 cells/cm^3. Reductions in bulk density from 82 to 89 percent were achieved. Average cell size exhibited a minimum with increasing pressure (or CO_2 density), and cell number density and foam bulk density exhibited corresponding maxima with increasing pressure (or CO_2 density). Possible reasons for this behavior include cell coagulation due to low melt strength and low interfacial tension at high pressure, or a situation where the phase diagram of the polymer-CO_2 mixture exhibits a maximum in swelling with increasing pressure.

Both tissue engineering and guided tissue regeneration (GTR) require the use of porous biodegradable polymeric materials, although each application has distinct requirements. In tissue engineering, highly porous macrocellular foams are used as scaffolds for cell seeding in vitro and subsequent implantation in vivo. The foams serve as temporary supports for cell growth and degrade over time as the cells become established [1-3]. In guided tissue regeneration, microporous implants are used as size selective membranes to promote the growth of a specific tissue at a wound site (e.g. bone regeneration) [4-8]. The membrane allows nutrients and wastes to permeate, while excluding the migration of undesirable soft tissue into the site. Foams having cell sizes less than 10 microns, or microcellular foams, would be required for guided tissue regeneration as it applies to bone regeneration. Further, the implant should be inherently biocompatible, have well-defined cell sizes, be resorbable with appropriate biodegradation rates, and be manageable in a surgical setting. Poly(lactic-co-glycolic acid) (PLGA) is widely-used in in-vivo applications as it biodegrades to relatively benign lactic and glycolic acid, and the degradation rate of the copolymer can be controlled by varying the ratio of the comonomers. The Food and Drug Administration (FDA) has approved its use in other in vivo applications, e.g. resorbable sutures [9]. Porous PLGA has been used extensively in recent work on cell culturing for later implantation.

Microcellular polymeric materials are often produced via phase separation of a homogeneous solution, followed by removal of the solvent [10-12]. The phase separation can be induced by a temperature quench or by the addition of a non-solvent to an initially homogeneous solution. The solvent is typically removed by freeze drying to protect the delicate pore structure from collapse induced by capillary forces associated with a retreating liquid-vapor meniscus. Microcellular foams can also be generated by the inorganic aerogel preparation route, where an infinite network is formed via a condensation polymerization that locks in the microcellular structure, followed by careful removal of the solvent [13]. Another method for generating porous membranes is a solvent-casting, particulate-leaching process [14], where the polymer is dissolved in an organic solvent and insoluble particulates are subsequently added. Evaporation of the solvent followed by removal of the particulates via leaching with a solvent (good solvent for the particles, non-solvent for the polymer) results in a foam with pores sizes equal to the size of the particulate matter. PLGA foams have been produced by this method using salt as the particulate. Each of the above processes involve the use of organic solvents, and unfortunately, residual solvent in the foam can damage cells causing potential problems for use in tissue engineering applications.

A method for producing microcellular foams without the use of conventional organic solvents employs supercritical fluid swelling followed by a pressure quench [15]. Here, the polymer absorbs an amount of fluid sufficient to lower its T_g well below the processing temperature, producing a homogeneous liquid polymer solution. A rapid quench in pressure causes the generation of nuclei due to supersaturation, and the cells grow until the polymer vitrifies and the cellular structure is locked in. Goel and Beckman produced poly(methylmethacrylate) foams using supercritical carbon dioxide which had average cell sizes in the range of 0.4 to 20 μm and foam bulk densities in the range of 0.4 to 0.9 g/cm^3 [16,17]. Here it must be remembered that the liquid-liquid phase envelope for a polymer-CO_2 mixture will be a closed loop which intersects the y-axes (both at 0% and 100% polymer) at the vapor pressures of CO_2 and the the polymer (essentially zero). In our two-phase system, we have a polymer highly swollen by CO_2 (a homogeneous solution) in equilibrium with what is essentially pure CO_2 (likely some very small amount of polymer dissolved). The fact that the swollen polymer is quite concentrated in polymer does not make it any less of a homogenous solution than a mixture dilute in polymer.

More recently, Mooney and coworkers [18] used this swell-quench method to generate macroporous PLGA using gaseous carbon dioxide. The PLGA was swollen by CO_2 at room temperature and 800 psi for 72 hours, and the resulting foams exhibited cell sizes in the 100-500 μm range with 94% void volume.

In our investigation, we generated microcellular PLGA using supercritical carbon dioxide. The effect of pressure and temperature on average cell size, foam bulk density, and cell number density were studied.

Experimental

Pellets of 85:15 lactide-glycolide copolymer (Medisorb Technologies, $T_g=45°C$) were first formed into disks via vacuum compression molding at 100°C for 15 minutes. The disks were 11 mm in diameter, 2.5 mm thick and weighed 0.30 grams.

The disks were foamed in a stainless steel high pressure cell equipped with a pressure transducer, thermocouple, and heating jacket. The temperature of the cell was controlled to within ± 1°C by a PARR Model 4842 PID temperature controller. A Haskel air driven gas booster (compressor) was used to pressurize the carbon dioxide to fill the cell. In a typical experiment, the disk was placed into a Pyrex liner and placed into the high pressure cell. The cell was flushed three times with carbon dioxide and then pressurized and heated to the desired saturation pressure and temperature. Exposure times varied from 20 minutes to 24 hours. Once the experiment was completed, the pressure was quenched within 30 seconds via venting of the carbon dioxide. Bulk densities of the foamed samples were determined via simple liquid displacement measurements.

Scanning electron micrographs were taken of each sample at 5 kV and magnifications ranging from 100x to 500x. The samples were prepared for SEM by freezing in liquid nitrogen, fracturing the surface, mounting the fracture fragment on stubs with carbon paint, and sputter coating with a 100 Å layer of gold. The SEM images were used for the determination of average cell size and size distribution. A LECO model 2001 image analyzer was used to generate a binary image from which the mean cell size for each cell is calculated from eight diameter measurements. A histogram was generated to determine the average cell size and standard deviation for the foam sample.

For foams produced at 40°C, data regarding cell size and bulk density represent averages from three to five different foams, while data for foams generated at 35°C represent averages for two foams.

Results

In order to determine the amount of time necessary to reach equilibrium in this system, experiments at 40°C and 5000 psi were run while varying the exposure time from 20 minutes to 24 hours. The average cell size at the center of the foam sample was measured and studied as a function of time. All of the experiments with exposure times of two hours or greater had the same average cell size (within experimental error). Therefore, all subsequent experiments were held at pressure for at least 4 hours to ensure that equilibrium swelling was achieved.

Figure 1a shows a scanning electron micrograph of foamed PLGA illustrating the typical structure of the cells generated via the pressure quench method. The cells have an open structure which can be desirable in GTR and other tissue engineering applications. This particular sample was run at 40°C and 5000 psi for 4 hours, and exhibits an average cell size of 11.2 μm and a bulk density of 0.26 g/cm^3 (the bulk density of PLGA is 1.4 g/cm^3). Figure 1b shows the type of distribution of cell sizes which were typically observed.

Figure 2a shows the effect of pressure on average cell size at 35°C, where cell size decreases with increasing pressure until a minimum size is reached at approximately 3000 psi. In Figure 2b, it appears that a maximum in foam bulk density occurs at approximately the same pressure. Given that the original density of the polymer is 1.4 g/cm^3, reductions in density of 82 to 89 percent were achieved, depending on the saturation conditions. The cell number density, or the number of cells per volume of foam, can be calculated from foam bulk density and number-average cell size via the following expression:

$$N = \frac{\text{number of cells in foam}}{\text{foam volume}} = \left[\frac{\left(\dfrac{\text{void space}}{\text{average cell volume}}\right)}{\text{foam volume}}\right] = \frac{\left[\left(\dfrac{m_p + m_g}{\rho}\right) - \left(\dfrac{m_p}{\rho_o}\right)\right]}{\dfrac{4}{3}\pi R_{avg}{}^3 F_1 F_2}$$

$$\left(\dfrac{m_p + m_g}{\rho}\right)$$

where m_p and m_g are the mass of the polymer and the mass of the gas, respectively, ρ and ρ_o are the final foam density and the initial polymer density, respectively, and R_{avg} is the number average radius of the cells . The factors F_1 and F_2 include the influence of a distribution of cell sizes on the calculation of the total void volume, where $F_1 = R_A/R_{avg}$, and $F_2 = R_V/R_{avg}$, and R_A and R_V are the "area" and "volume-average" cell sizes:

$$R_A = \frac{\Sigma\left(N_i R_i{}^2\right)}{\Sigma\left(N_i R_i\right)} \qquad \text{and} \qquad R_V = \frac{\Sigma\left(N_i R_i{}^3\right)}{\Sigma\left(N_i R_i{}^2\right)}$$

Assuming that the mass of the gas is negligible compared to the mass of the polymer, the cell number density becomes a function of average cell size and foam bulk density:

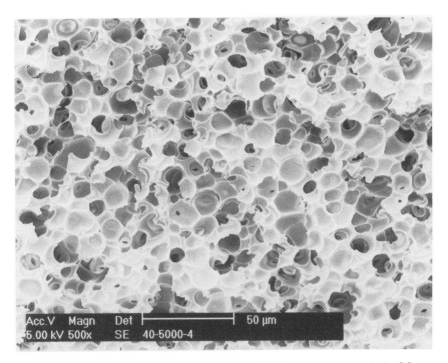

Figure 1a. Typical microcellular structure of PLGA foam generated via CO_2 pressure quench (T = 40°C, P = 5000 psi).

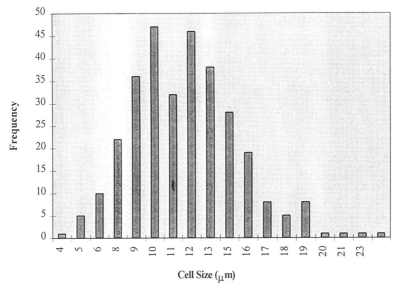

Figure 1b: Distribution of Cell Size (PLGA-CO_2, T=40°C, P=5000 psi)

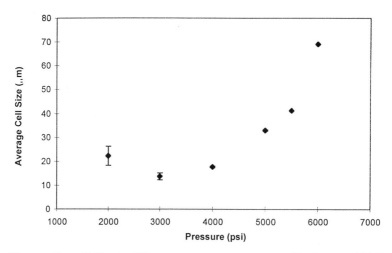

Figure 2a: Effect of Pressure on average cell size at 35°C

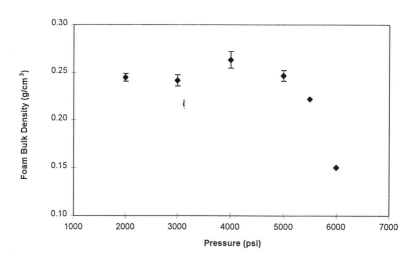

Figure 2b: Effect of Pressure on foam bulk density at 35°C

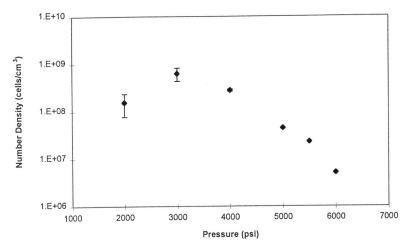

Figure 2c: Effect of Pressure on cell number density at 35°C

$$N = \left[\frac{\left(1 - \dfrac{\rho}{\rho_o}\right)}{\dfrac{4}{3}\pi R_{avg}^{3} F_1 F_2} \right]$$

Further, examination of the histograms for the various foams showed that F_1 and F_2 average 1.08 and 1.15, respectively, and that variations in temperature or pressure did not lead to significant variations in either F_1 or F_2. Thus, cell number density was calculated using the above equation; Figure 2c shows cell number densities comparable to those observed in other microcellular polymeric foams produced using CO_2.

Figures 3a and 3b show the effect of pressure at various temperatures, where only a minimal effect of temperature in the range studied is observed. At each temperature the same trends in the data versus pressure is observed, but the various curves are shifted horizontally with increasing temperature, suggesting that CO_2 density governs behavior. The data were therefore plotted as a function of CO_2 density (Figures 4a and 4b); the fact that the curves superimpose suggest that CO_2 density does indeed dominate the foaming results.

Discussion

Upon prolonged exposure to CO_2 at high pressure, PLGA absorbs a sufficient amount of supercritical CO_2 to lower its T_g well below the processing temperature, producing a homogeneous liquid polymer solution. High pressure DSC experiments performed in our lab suggest that the T_g of PLGA is lowered from 45°C to below 35°C after exposure to CO_2 at only 300 psi at 35°C. For the case of CO_2-generated PMMA foams, Goel and Beckman showed that classical nucleation theory could be used to describe nucleation activity during the swell-quench process. This is not surprising, because the polymer is plasticized to the point where a homogeneous liquid exists at the operating temperature. A rapid quench in pressure causes the generation of nuclei due to supersaturation, and the cells grow until the polymer vitrifies and locks in the cellular structure. According to classical nucleation theory, the homogeneous nucleation rate, or N^0_{homo}, is

$$N^0_{homo} = C_0 \left(\frac{2\gamma}{\pi m B} \right)^{1/2} \exp\left(\frac{-\Delta G_{homo}}{kT} \right)$$

where C_0 is the gas concentration, m is the mass of a molecule, and B is the frequency factor. ΔG_{homo} is the energy barrier for nucleation and can be calculated as a function of the interfacial tension of the binary mixture, γ, and the magnitude of the pressure quench, ΔP:

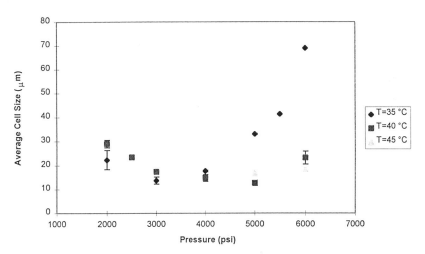

Figure 3a: Effect of pressure and temperature on average cell size

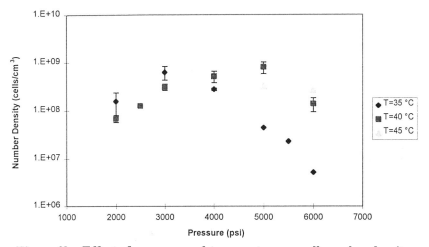

Figure 3b: Effect of pressure and temperature on cell number density

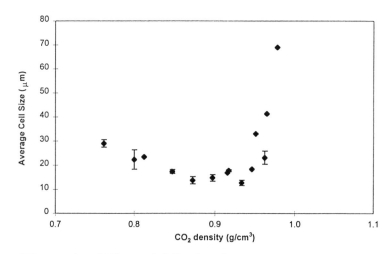

Figure 4a: Effect of CO_2 density on average cell size

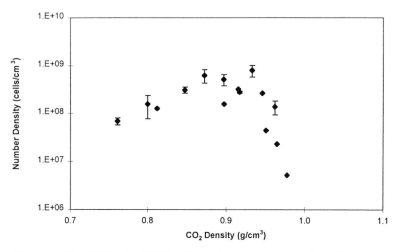

Figure 4b: Effect of CO_2 density on cell number density

$$\Delta G_{homo} = \frac{16\pi\gamma^3}{3\Delta P^2}$$

As pressure increases, the amount of CO_2 absorbed by the polymer increases, resulting in an increase in CO_2 concentration (C_0), a decrease in mixture interfacial tension, γ [19], and, of course, an increase in the magnitude of the quench (ΔP). The ultimate result is an increase in nucleation rate. However, at higher pressures, extensive swelling can depress the interfacial tension to very low levels, and thus the rate of bubble nucleation could be governed by the prefactor rather than the exponential term (which will tend towards unity). Consequently, if the interfacial tension decreases faster than the degree of swelling increases, the rate of nucleation could drop at high pressures, although the drop should be gradual. In CO_2-generated PMMA foams, Goel and Beckman showed that the nucleation rate, and thus the total number of cells generated during the pressure quench, increased with increasing pressure up to 5000 psi, leveling off thereafter. In our system, however, the cell number density exhibits a maximum value and then decreases with increasing pressure. Possible reasons for this behavior will be discussed later in this paper.

In general, the cell sizes produced in the PLGA foams (10 to 70 μm) were larger than those produced in PMMA foams (0.4 to 20 μm) generated by Goel using CO_2. Goel developed a model to predict cell sizes in the PMMA/CO_2 system, accounting for the effect of pressure (and thus CO_2 concentration in the polymer) on the critical bubble radius ($r_c = 2\gamma/\Delta P$), and the various physical properties governing bubble growth (viscosity, diffusivity, etc.). In general, Goel found that the effect of CO_2 concentration on the interfacial tension of the polymer-gas mixture, and thus its effect on the initial bubble radius, tended to dominate the growth simulation.

In addition to interfacial tension, the pressure (P_g) at which the porous structure vitrified due to efflux of CO_2 from the polymer during the pressure quench also exerted an important influence on the ultimate average cell size. Bubbles will grow until the system reaches a pressure, P_g, at which sufficient CO_2 has left the matrix to raise the T_g to a point equal to or above ambient temperature, where the polymer then vitrifies. At higher pressures, growth is due primarily to the diffusion of CO_2 molecules from the polymer matrix into the pores. At lower pressures, however, the growth is also driven by expansion of the gas (the molar volume of CO_2 increases more rapidly with decreasing pressure at lower absolute pressures, given that the compressibility of CO_2 increases rapidly as pressure drops below 1000 psi). For the PMMA/CO_2 system, P_g is apparently high enough to lock in the porous structure formed via the pressure quench before significant gas expansion occurs. The apparent lower P_g for the PLGA/CO_2 system is due in large part to the lower T_g^0 of PLGA (45°C vs. 105°C for PMMA). It is likely that bubble growth in the PLGA/CO_2 system is not only driven by diffusion, but also by gas expansion, leading to larger cells. Goel's growth model predicts that cell size will decrease with increasing pressure, largely due to decreases in the size of the critical nucleus, itself indirectly a function of the amount of CO_2 absorbed in the polymer. Experimental data for PMMA/CO_2 (pressures up to 5000 psi) verified the model predictions. However, the experimental data for the PLGA/CO_2 system suggest that cell size decreases, reaches a minimum value, and begins to increase again with increasing pressure (or CO_2 density). The cell number exhibits a corresponding maximum, falling at higher pressures. One possible explanation for the min/max behavior in both cell size and number could be due to the coalescence of bubbles at higher pressures. At higher pressures, both the viscosity and interfacial tension of the polymer could be so greatly suppressed due to extensive swelling by CO_2 that the bubbles would not survive collisions with each other under these conditions, leading to smaller numbers of larger bubbles.

Another possible, but less likely explanation for the min/max behavior could be the somewhat non-intuitive result that the weight fraction of CO_2 absorbed by the polymer decreases at higher pressures. As discussed above, nucleation rate is a strong function of CO_2 concentration and mixture surface tension, which are both functions of the weight fraction of CO_2 (w_{CO2}) in the polymer mixture. If w_{CO2} decreased, the nucleation rate would decrease, resulting in fewer cells. Goel's model suggested that ultimate cell size is a strong function of critical nucleus radius, which is in turn a function of mixture surface tension. If w_{CO2} decreased, then the surface tension would increase and the critical nucleus radius would increase. The initial size of each bubble would be larger, possibly resulting in larger final bubbles. Therefore, if at higher pressures the amount of CO_2 absorbed by the polymer decreases, one would expect fewer, larger bubbles. A decrease in the extent of swelling as pressure increases at high pressures suggests an "hour-glass" type phase diagram in P-x space for the PLGA-CO_2 mixture. Although such behavior has not been observed previously in polymer-CO_2 systems, almost no data exists on the swelling behavior of amorphous polymers at extremely high pressures.

Conclusions

PLGA foams were produced by a pressure quench method employing supercritical carbon dioxide. In general, the cell sizes produced were larger than in other foams, i.e. PMMA, produced via this method. This is due to the fact the PLGA is more plasticized by the carbon dioxide owing in part to its lower T_g^0. The foams had average cell sizes ranging from 10 to 70 μm, with cell number densities from 10^6 to 10^9 cells/cm^3. Reductions in bulk density from 82 to 89 percent were achieved. Average cell size exhibited a minimum with increasing pressure (or CO_2 density). Cell number density and foam bulk density exhibited maximums with increasing pressure (or CO_2 density). The minimum and maximum values occurred at a CO_2 density of approximately 0.9 g/cm^3.

Acknowledgment

DS acknowledges the financial support of the National Science Foundation through a Graduate Training Grant to the Department of Chemical Engineering at the University of Pittsburgh.

REFERENCES

1. A.G. Mikos, G. Sarakinos, et. al., *Biotechnol Bioeng*, **42**, 716 (1993)
2. D.J. Mooney, S. Park, et. al., *J. Biomed. Mater. Res.*, **29**, 959 (1995)
3. J.P. Vacanti, M.A. Morse, et. al., J. Pediatr. Surg., **23**, 3
4. N.M. Blumenthal, *J Periodontol*, **59**, 830 (1988)
5. C. Dahlin, L. Sennerby, et. al., *Int J Oral Maxillofac Implant*, **4**, 19 (1989)
6. A. Linde, P. Alberius, et. al., *J Periodontol*, **64**, 1116 (1993)
7. T. Scantlebury, *J Periodontol*, **64**, 1129 (1993)
8. J. Gotlow, *J Periodontol*, **64**, 1157 (1993)
9. E.J. Frazza and E.E. Schmitt, *J. Biomed. Mater. Res.*, **1**, 43 (1971)
10. A.T. Young, *J. Cell. Plast.*, **23**, 55 (1987)
11. D.R. Lloyd and J.W. Barlow, in "New Membrane Materials and Processes for Separation", K.K. Sirkar and D.R. Lloyd, eds., AIChE Symp. Ser., New York (1988)
12. J.H. Aubert and A.P. Sylwester, *Chemtech*, **May**, 290 (1991)
13. J. Fricke, ed., "Aerogels", Springer-Verlag, Berlin FRG (1986)
14. A.G. Mikos, G. Sarakinos, et. al., *Biomaterials*, **14**, No.5 (1993)
15. S.K. Goel and E.J. Beckman, "Generation of Microcellular Polymers using Supercritical CO_2",in *Cellular Polymers*, Vol. 12., J.M. Buist, ed., Rapra Technology Ltd., UK (1993)
16. S.K. Goel and E.J. Beckman, *Polym Eng Sci*, **34**, 1137 (1994)
17. S.K. Goel and E.J. Beckman, *Polym. Eng. Sci.*, **34**, 1148 (1994)
18. D.J. Mooney, D.F. Baldwin, et. al., *Biomaterials*, **17**, 1417 (1996)
19. A.V. Yazdi and E. J. Beckman, "Bubble Nucleation in Polymer Mixtures," in *Polymer Devolatilization*, Marcel Dekker, New York (1996)
S.K. Goel and E.J. Beckman, *AIChE Journal*, **41**, No. 2, 357 (1995)

POLYMER NETWORKS

Chapter 12

Polysulfone Oligomers with Thermosetting End-Groups and Their Curing Behavior Under Solvent-Free Conditions

Rumiko Hayase and Tetsuo Okuyama

Materials and Devices Laboratories, Research and Development Center, Toshiba Corporation, 1, Komukai Toshiba-cho, Saiwai-ku, Kawasaki 210-8582, Japan

Abstract

Polysulfone oligomers with thermosetting terminals were synthesized and their crosslinked materials were investigated. The thermal curing of polysulfone oligomers with styryl and propargyl end-groups produced thermally stable crosslinked materials. The melt-viscosity of the polysulfone compounds can be modified by choosing the main chain, the end-group, the molecular weight and the content of inorganic fillers. The water absorption of the crosslinked oligomers was lower than that of conventional epoxy resins. The cured compositions containing more than 70wt% of silica had flame resistance corresponding to the UL94 V-0 without additional flame-retardants. They could be candidates for use as encapsulation resins for semiconductors and printed circuit boards.

Introduction

Semiconductors are molded by thermosetting resins in order to protect the chip from mechanical destruction and electrical failure caused by moisture (Figure 1). Typical epoxy encapsulation compounds are composed of epoxy resins, phenol resins, curing catalysts, silicas, flame-retardants, pigments and so forth (Figure 2). Epoxy resins have been employed extensively, due to their excellent thermal stability, adhesion strength, electrical insulation and ease of processing (Table 1)(1). However, the water absorption is high, and the cured resins are brittle because of the highly crosslinked structure. Flame-retardants such as antimony oxide or halogen compounds are required in order to achieve flame-resistance, but it is desirable to avoid their use due to environmental considerations.

In view of these issues, at the research level there is a growing tendency to use engineering plastics (2). For example, PPS (poly-phenylene-sulfide) has high thermal stability, low water absorption and high flame-resistance (Table 2). However, their high melt-viscosity inhibits the use of engineering plastics for microelectronics packages, because thin bond wires are deformed during the molding process.

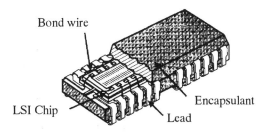

Figure 1. Plastic encapsulated microelectronics.

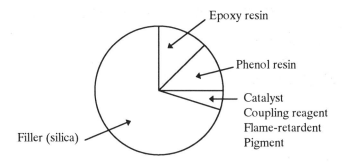

Figure 2. Epoxy encapsulant.

In order to overcome these disadvantages, we synthesized functionalized oligomers shown in Figure 3. We made the thermoplastic chain length shorter and introduced the thermosetting moieties on the ends. Therefore, the viscosity is low before curing, and thermosetting end-groups are crosslinked to give thermally stable products after molding. Our goals for encapsulation compounds are summarized in Table 3. Namely, melt-viscosity is less than 50 Pa·S, gel time is lower than 60 seconds, water absorption is lower than conventional epoxy encapsulants, coefficient of thermal expansion is lower than 20 ppm/K, and flame-resistance is V-0 grade of UL94 standard without flame-retardants. In addition, the compounds should not have volatile organic compounds.

It has been reported that thermal curing of polysulfones with styryl end-groups (*3*) and dipropargyl ethers of bisphenols (*4*) produces highly crosslinked networks. We synthesized polysulfone oligomers with thermosetting terminals to obtain novel resin compositions for electronic applications. In this paper, their curing behavior, thermal stability, viscosity, water absorption and flammability are discussed.

Table 1. Epoxy encapsulants.

Advantages	Disadvantages

☆ High thermal stability
☆ High adhesion strength
☆ Excellent electrical insulation
☆ Easy processing
	★ High water absorption
	★ Brittle due to crosslinking
	★ Flammable
	  Flame-retardant is necessary.
	  (halogen compounds, antimony oxide)
	  → Environmental problems

Table 2. Thermoplastic encapsulants. (engineering plastics).

Advantages	Disadvantages
☆ High thermal stability ☆ Low water absorption ☆ High flame-resistance	★ High melt-viscosity → Deformation of bonding wires

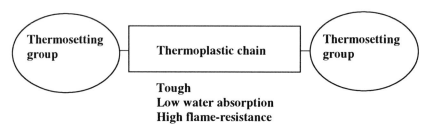

Tough
Low water absorption
High flame-resistance

Figure 3. How to cope with disadvantages for application to semiconductor encapusulants.

Table 3. Requirement.

(1) Melt-viscosity	< 50 Pa·S
(2) Gel-time	< 60 sec
(3) Water absorption	Lower than conventional epoxy encapsulants
(4) Coefficient of thermal expansion	< 20 ppm / K
(5) Flame-resistance	UL94 V-0, without flame-retardants

Experimental

Polysulfone oligomers

The synthesized oligomers are summarized in Figure 4. All syntheses were conducted in nitrogen atmosphere. The molar amount of employed bisphenols was higher than that of dichlorodiphenyl sulfones. In order to control molecular weight, the molar ratio of bisphenols and dichlorodiphenyl sulfone was modified. The structures of the synthesized oligomers were determined by ^{1}H-NMR. The molecular weights of the oligomers were estimated by gel-permeation chromatography.

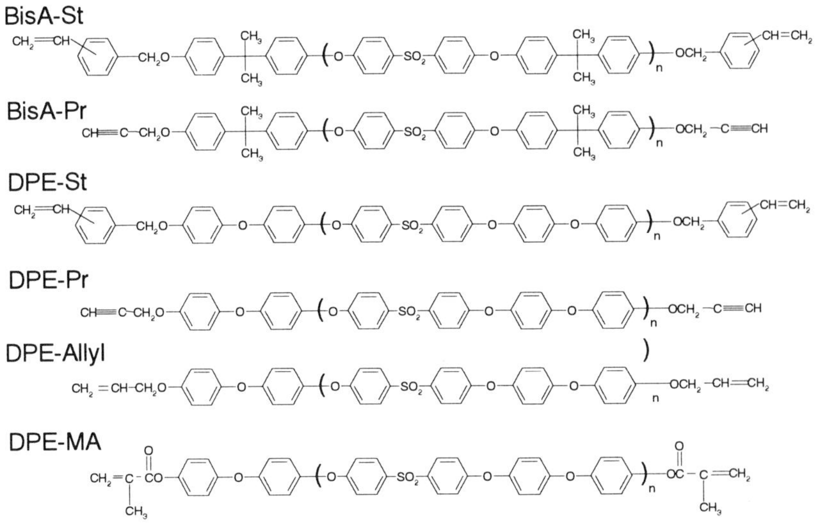

BisA-St

BisA-Pr

DPE-St

DPE-Pr

DPE-Allyl

DPE-MA

Figure 4. Polysulfone oligomers with thermosetting end groups.

BisA-St (Bisphenol A polysulfone with styryl terminal)

Bisphenol A, NaOH aq. and toluene were added to N-methylpyrrolidone. The solution was heated with stirring for 2 hours at 130-150°C while azeotropically removing water with toluene, and thereby obtaining sodium salt of bisphenol A. To this reaction solution was added 4,4'-dichlorodiphenyl sulfone. The resultant solution was heated at 150-160°C with stirring for 12 hours while azeotropically removing water with toluene, and then toluene was distilled off. After cooling the reaction solution down to room temperature, vinylbenzyl chloride (m,p-mixture) was added to the reaction solution. The solution was heated at 60°C with stirring for 10 hours. After the reaction solution was diluted by the addition of dichloromethane, salts were filtered off. The filtrate was washed with 1wt% of oxalic acid aq., and then with water. The solution was dried with $MgSO_2$. After the solution was concentrated, it was poured into methanol, and precipitated oligomer was collected through filtration. Subsequently, the oligomer was dissolved in dichloromethane and precipitated in methanol again. The oligomer was collected and dried in a vacuum oven (Figure 5).

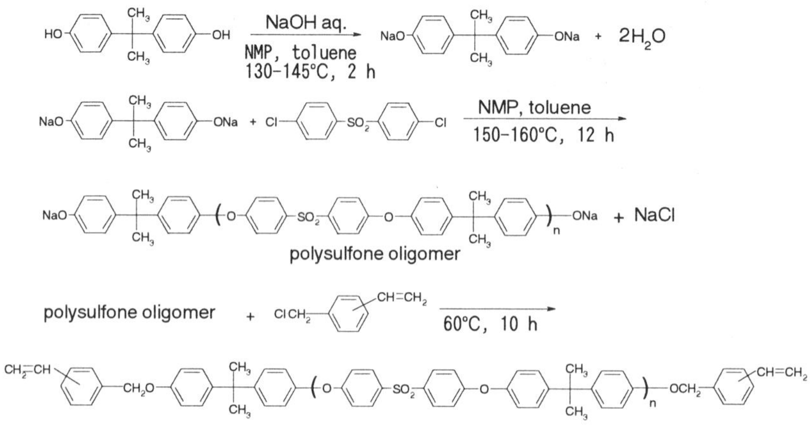

Figure 5. Synthesis of polysulfone oligomers with thermosetting end groups.

BisA-Pr (Bisphenol A polysulfone with propargyl terminal)
   BisA-Pr was synthesized in the same manner as BisA-St except for the use of propargyl bromide instead of vinylbenzyl chloride.
DPE-St (Diphenyl ether polysulfone with styryl terminal)
   DPE-St was synthesized in the same manner as BisA-St except for the use of 4,4'-dihydroxydiphenyl ether instead of bisphenol A.
DPE-Pr (Diphenyl ether polysulfone with propargyl terminal)
   DPE-Pr was synthesized in the same manner as BisA-Pr except for the use of 4,4'-dihydroxydiphenyl ether instead of bisphenol A.
DPE-Allyl (Diphenyl ether polysulfone with allyl terminal)
   DPE-Allyl was synthesized in the same manner as DPE-St except for the use of allyl bromide instead of vinylbenzyl chloride.
DPE-MA (Diphenyl ether polysulfone with methacryl terminal)
   DPE-Allyl was synthesized in the same manner as DPE-St except for the use of methacryloyl chloride instead of vinylbenzyl chloride.
Filler   Fused silica. The average diameter was 22μm.
Thermal analysis
   DSC and TGA were measured by Seiko Instruments SSC/5200 DSC 220C and TG/DTA 220, respectively. All measurements were carried out in N_2 atmosphere.

The oligomers were heated to 320°C for the 1st scan of DSC analysis at the rate of 5°C/min, and then cooled to 20°C at the rate of 50°C/min. The glass-transition temperature of cured oligomers was estimated by the 2nd scan of DSC analysis. TGA was measured at the rate of 10°C/min.

Gel-time

The silica filled oligomers were pressed at room temperature to form disks with 1 mm thickness and 13 mm diameter. The gel-time was evaluated by visco-elastic measurement. The frequency was 3Hz.

Melt-viscosity

The melt-viscosity of the oligomers and their compounds was measured by a capillary rheometer (Shimadzu Flowtester, FT500A). The nozzle diameter was 1mm, the load was 10Kgf/cm^2.

Water absorption

The silica filled oligomers were pressed at 180°C to form disks with 1 mm thickness and 30 mm diameter, and then cured at 250°C for 2 hours. After the disks were dried in an oven at 150°C for 24 hours, they were placed in an oven at 85°C and 85% humidity for 168 hours. The water absorption was calculated from the weight difference.

Thermal expansion

The coefficient of thermal expansion of the cured oligomers was measured by Seiko Instruments TMA/SS120C.

Flame resistance

The flammability was examined according to the UL94 standard.

Results and discussion

The characteristics of the synthesized polysulfone oligomers are shown in Table 4.

Table 4. Polysulfone oligomers.

	$\overline{Mn}$	$\overline{Mw}/\overline{Mn}$	Yield of oligomer (%)	Thermosetting group/End-group
BisA-St	1250	1.53	86	0.80
BisA-Pr (a)	1410	1.39	89	0.85
BisA-Pr (b)	2320	1.50	85	0.76
DPE-St (a)	1130	1.45	80	0.80
DPE-St (b)	2390	1.67	85	0.75
DPE-Pr	2400	1.79	78	0.60
DPE-Allyl	2380	1.78	79	0.60
DPE-MA	2630	1.71	82	0.40

The molar ratio of employed bisphenols and diphenyl dichlorosulfone for the syntheses of BisA-St, BisA-Pr (a) and DPE-St (a) was 1.5 : 1.0. It was 1.3:1.0 for BisA-Pr (b), and 1.1:1.0 for DPE-St (b), DPE-Pr, DPE-Allyl and DPE-MA, respectively. The yields of the oligomers were around 80 - 90 %, and the ratios of introduced thermosetting end-groups were 40 - 85 %. The flow temperature of oligomers depended on the molecular weight and the structure of the polysulfone backbone. Oligomers with higher molecular weight melted at higher temperature. DPE-St (a) melted at lower temperature than BisA-St (a) which had comparable molecular weight, because the ether in the main chain of DPE-St is more flexible than isopropylidene of BisA-St. The polysulfone oligomers with styryl, propargyl and allyl end-groups were thermally cured (Table 5).

Table 5. Characteristics of cured polysulfone oligomers.

	$\overline{Mn}$	flow temp.(°C)	cure start (°C)	exothermic peak(°C)	Tg of cured oligomers	5wt% weight loss temp (°C)
BisA-St	1250	110-120	169	214	145	407
BisA-Pr (a)	1410	115-125	201	270	163	428
BisA-Pr (b)	2320	150-170	201	269	171	431
DPE-St (a)	1130	90-100	170	212	141	413
DPE-St (b)	2390	130-140	172	228	167	432
DPE-Pr	2400	135-145	200	267	176	431
DPE-Allyl	2380	125-135	166	248	150	441
DPE-MA	2630	145-160	-	-	175 (melt)	433

The thermosetting temperature depended on the end-groups. Both styryl oligomers and allyl oligomers started crosslinking at about 170 °C (Table 5). However, the exothermic peak temperature of styryl oligomers was lower than that of allyl oligomer, which explains why the crosslinking speed of styryl oligomer was higher than that of allyl oligomer. The cure-start temperature of propargyl oligomers was 200°C, which was higher than that of styryl and allyl oligomers. On the other hand, DPE-MA with methacryl end-group had no exothermic peak and did not crosslink thermally. The low ratio of methacryl groups may have contributed to reduction of the curing rate of DPE-MA. Because phenols are effective as inhibitors for radical polymerization, the residual phenol end-groups may inhibit the crosslinking reaction of methacryl group. In order to accelerate the crosslinking reaction, the ratio of thermosetting end-groups should be increased and the phenols must be decreased. There is a possibility that Michael-addition may take place for the methacryl groups. The reaction progresses via carbanion intermediates which are activated by polar solvent or active hydrogen. However, in this study, oligomers were heated under solvent-free conditions, and the relative amount of phenol end-groups should be much less than that of polar solvents which were used for solvent polymerization, and no catalysts were added which ionize the methacryl groups. Therefore, the reaction may not have occurred. However, because the melting temperature of DPE-MA was raised after heating, it should be crosslinked partly.

The crosslinked polysulfones had a high Tg and thermal decomposition temperature. The glass-transition temperatures of the cured oligomers depended on the thermosetting groups. The glass-transition temperatures of the crosslinked DPE-Pr, DPE-St (b) and DPE-Allyl were 176°C, 167°C and 150°C, respectively. Higher molecular weight oligomers had higher thermal decomposition temperatures. This may result from the polysulfone main chain being more thermally stable than the thermosetting end-groups. The 5% weight loss temperatures of these polysulfone oligomers were higher than 400°C, indicating excellent thermal resistance.

The isothermal viscosity of oligomers with 70 wt% of silica was measured to observe their thermosetting behavior (Figure 6).

Because of the difference of the thermosetting temperature for the oligomers, the viscosity was measured at 170°C for BisA-St and at 250°C for BisA-Pr (b). BisA-St crosslinked within 500 sec. The gel-time of BisA-Pr (b) was around 5000 sec, and much longer than that of BisA-St. The viscosity slope of BisA-Pr (b) changed at about 2500 sec, which suggests propargyl groups crosslinked via two different reactions. BisA-St did not have the corresponding inflection point. Dirlikov has reported that dipropargyl ether of bisphenol A was B-staged at 185°C, and then cured by an unknown mechanism (4). The viscosity curve of BisA-Pr (b) may reflect these curing characteristics.

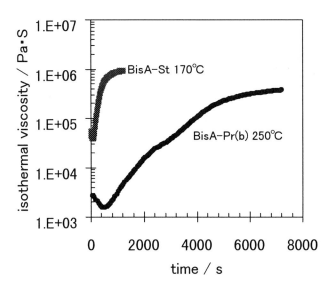

Figure 6. The isothermal viscosity of BisA-St at 170°C and BisA-Pr(b) at 250°C. Filler content: 70 wt%.

The gel-times of silica-filled oligomers were evaluated by the visco-elastic measurement (Table 6). The gel-times of oligomers themselves were almost the same as those of silica filled compounds. For commercial use, the required gel-time is less than 60 seconds; however, the gel-time of oligomers was much longer than the requirement. In order to shorten the gel-times, 2wt% of dicumyl peroxide was added as a curing catalyst, with an additional 10 wt% of 2,2-bis[4-(4-maleimide phenoxy)phenyl] propane added as a curing accelerator. The gel-times decreased to between 10 - 70 sec, which almost satisfies the requirement. The curing reaction of propargyl oligomers did not accelerate sufficiently only with adding of dicumylperoxide or bismaleimide. Their combination may facilitate a cooperative free-radical crosslinking reaction.

Table 6. Gel-time.

	without catalyst	with catalyst
BisA-St	470sec (170°C)	70 (170°C)
DPE-St (a)	1480 (170°C)	40 (170°C)
BisA-Pr (a)	>4000 (250°C)	< 10 (250°C)
BisA-Pr (b)	4830 (250°C)	< 10 (250°C)

filler content : 70 wt%.
curing catalyst : 2 wt% of dicumyl peroxide was added, with an additional 10 wt% of bismaleimide added as a curing accelerator.

When the flowability of the sealing resin decreases, bonding wires of LSI chip are deformed. Therefore, the melt-viscosity is one of the most important factors for encapsulants. The melt-viscosities were measured at 170°C for styryl oligomers, and at 250°C for propagyl oligomers (Table 7). The viscosity of the DPE-St (a) was lower than that of BisA-St, because DPE-St (a) has a more flexible main chain than BisA-St. Usually, when the processing temperature increases, the melt-viscosity of polymers decreases. However, the viscosity of BisA-St was not reduced sufficiently by raising the temperature, since BisA-St crosslinked rapidly above 180°C. The thermosetting temperature of propargyl oligomers was higher than that of styryl oligomers, and accordingly, they can be processed at higher temperature to decrease their melt-viscosity. Except for BisA-St, melt-viscosities of these oligomers were around 30 Pa·S, and satisfied the target requirement.

The water absorption of the cured polysulfone compositions (filler content: 70 wt%) was much lower than that of conventional cresol novolac epoxy encapsulation compound (KE1000, Toshiba Chemical Co.) as shown in Figure 7. The water absorption of BisA-St was lower than that of DPE-St (a), probably because of the hydrophobicity of the isopropylidene chain structure of BisA-St.

Table 7. Melt-viscosity.

	filler	temp.	viscosity
BisA-St	70 wt%	170°C	300 Pa·sec
DPE-St (a)	70	170	32
BisA-Pr (a)	70	250	< 20
	80	250	32
BisA-Pr (b)	70	250	34

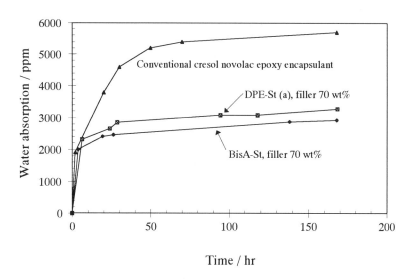

Time / hr

Figure 7.   Water absorption.   (humidity 85%, 85°C).

In commercial applications, inorganic fillers are added to the resins in order to match the coefficient of thermal expansion (CTE) of the encapsulants to the LSI chips or lead frames. The CTE of conventional epoxy encapsulants is around 10 to 20 ppm/K. The thermal expansion of the polysulfone oligomers can be modified by adding 70 to 80 wt % of silica (Figure 8).

When resins are used as encapsulants for semiconductors, flame-resistance (UL94 V-0 standard) is required. The flammability of crosslinked BisA-St and DPE-St was evaluated according to the UL94 standard. The cured oligomers which had no silica did not satisfy V-0 grade, and their flammability corresponded to V-2 class. However, for the compositions containing more than 70wt% of silica, their flame-resistance satisfied the UL94 V-0 standard without addition of flame-retardants (Table 8).

Conclusion

Thermal curing of polysulfone oligomers with styryl, propargyl and allyl end-groups produced thermally stable crosslinked materials. Styryl oligomers crosslinked faster than either propargyl or allyl oligomers. The melt-viscosity of oligomers can be reduced by choosing the structure of the main chain and end-groups. The water absorption of the crosslinked oligomers was lower than that of conventional epoxy resins. Cured compositions containing more than 70 wt% of silica had flame-

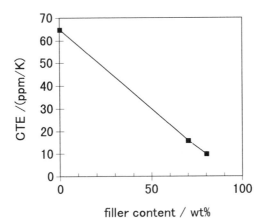

filler content / wt%

Figure 8. Dependence of coefficient of thermal expansion on filler content. Oligomer: DPE-St (a).

Table 8. Flammability of crosslinked oligomers.

	BisA-St		DPE-St (a)	
Filler content / wt%	70	80	70	80
UL94	V-0	V-0	V-0	V-0

resistance without the addition of flame-retardants such as antimony oxide or halogen compounds. Polysulfone oligomers with thermosetting end-groups are candidates for application as encapsulation resins for semiconductors.

References
(1) (a) R.R.Tummala, E.J.Rymaszewski, A.G.Klopfenstein, *Microelectroncs Packaging Handbook, Part II,* International Thomson Publishing, 1997.
(b) C.A.May, *Epoxy Materials, Electronic Materials Handbook, 1-Packaging,* ASM Intl., 1989.
(2) H.Sakai, M.Adachi, K.Suzuki, *Seikei-Kakou* **2**, (3), 220-226 (1990).
(3) J.L.Hedrick, J.G.Hilborn, R.B.Prime, J.W.Labadie, D.J.Dawson, T.P.Russel, V.Wakharker, *Polymer,* **35**, (2), 291-299 (1994).
(4) S.Dirlikov, *Polymer Preprints*, **35**, (1), 339-340 (1994).

Chapter 13

A New Route for Solid-State Cross-Linking in Miscible Polymer Blends and Organic–Inorganic Polymeric Hybrids

Eli M. Pearce, T. K. Kwei, and Shaoxiang Lu[1]

Department of Chemical Engineering, Chemistry and Materials Science, and the Herman F. Mark Polymer Research Institute, Polytechnic University, Six MetroTech Center, Brooklyn, NY 11201

A new route for solid state crosslinking in either binary miscible hydrogen-bonded polymer blends or organic-inorganic polymeric hybrids was developed by condensation of self-associated silanol groups. Miscible polymer blends or organic-inorganic polymeric hybrids were prepared by blending of a series of novel styrene- or inorganic siloxane-based silanol polymers with appropriate organic polymers. The miscibility in either polymer blends or polymeric hybrids was achieved by the formation of strong inter-polymer hydrogen bonds between the silanol groups and the hydrogen bond acceptor groups of counter polymers. The solid state crosslinking was conducted by subsequent condensation of self-associated silanol groups in either miscible blends or hybrids without the presence of any external crosslinkers or catalysts. This new route for solid state crosslinking by utilizing the dual nature of the silanol functional groups lead to the preparation of intimately mixed semi-interpenetrating polymer networks.

Interpenetrating polymer networks (IPNs) have been an important category of polymeric materials. There has been ample documentation in the literature on the subject (1,2). However, most IPNs and related materials investigated to date were phase separated. Only a limited number of examples of miscible IPNs were reported (3-5). In the hydrogen bonded polymer blends, it was demonstrated that controllable morphology can be achieved in the preparation of semi-interpenetrating polymer networks (semi-IPNs) (5-7). Miscible semi-IPNs were prepared from miscible blends of Novolac and poly(styrene-co-methyl methacrylate) in the presence of 1,3-dioxolane at 70 °C. The obtained miscible semi-IPNs showed a single phase up to

[1]Current address: Revlon Research Center, 2121 Route 27, Edison, NJ 08818

180 °C at the cloud point of blends. In contrast, when the cross linking reaction was carried out in the solid state above the cloud point 180 °C in the presence of hexamethylenetetramine, the obtained materials were heterogeneous and showed two T_g's. The formation of miscible or phase separated semi-IPNs in the hydrogen bonded polymer blends depended on whether the cross linking reaction is conducted above or below the lower critical solution temperature (LCST).

The silanol functional group has been long known as a reactive intermediate in silicon chemistry (8). The silanol groups produced in hydrolysis tend to condense with each other to form siloxane. Organosilanols are also known as stronger hydrogen bond donors than analogous alcohols due to d_π - p_π bonding between silicon and oxygen atoms that led to high polarity of O—H bond (9-16). In this paper, we describe a new route for solid state crosslinking in miscible hydrogen bonded polymer blends and organic-inorganic polymeric hybrids by utilizing the dual nature of silanol functional groups. Miscibility of blends or polymeric hybrids is achieved by the formation of strong inter-polymer hydrogen bonds between silanol groups and acceptor groups of counter polymers. On the other hand, chemical crosslinking via the formation of siloxane bonds is accomplished by the condensation of self-associated silanol groups in the solid state without the presence of any external crosslinkers or catalysts.

Experimental

Materials. Poly(methylhydrosiloxane) end capped with trimethylsilyl groups, with a reported molecular weight of 2270, was purchased from Hüls America Co. Organic polymers bearing different hydrogen bond acceptor groups for blending studies are shown in Table I. Poly(N-vinylpyrrolidone) (PVPr) and poly(4-vinylpyridine) (PVPy) were vacuum dried at 80 °C, poly(ethyl oxazoline) (PEOx) was vacuum dried at 60 °C for 3 days each and then stored in a desiccator before use.

Synthesis of inorganic siloxane-based silanol polymer. In a typical preparative procedure, 0.76 g poly(methylhydrosiloxane) (I) was dissolved in 10 ml acetone. To this solution a 275 ml dimethyldioxirane solution in acetone (18) (0.06 ~ 0.08 M) was added quickly and the reaction mixture was stirred at room temperature for 2 h. Poly(hydroxymethylsiloxane) (PHMS) (II) in acetone solution was obtained after concentrating the solution under reduced pressure at room temperature. Silanol polymer solution in isopropanol solution were prepared by addition of the solvent to the concentrated polymer solution in acetone followed by evaporation of the more volatile acetone solvent under reduced pressure. The process was repeated for at least three times to ensure maximum removal of acetone solvent. The resulting concentrated silanol polymer or copolymer solution in isopropanol was transferred into a volumetric flask and a 0.4% (g/ml) of polymer solution was prepared for solution blending.

Table I.   Organic Polymers Used in the Study

Polymer	Molecular Weight	T_g (°C)	Source
$-[CH_2-CH]_n-$ (Poly(N-vinylpyrrolidone) (PVPr))	Mw = 45,000	174	Polyscience, Inc.
$-[CH_2-CH]_n-$ (poly(4-vinylpyridine) (PVPy))	Mw = 40,000	150	Polyscience, Inc.
$-[N-CH_2-CH_2]_n-$ (poly(ethyl oxazoline) (PEOx))	Mw = 50,000	58	Dow Chemical Co.

Synthesis of styrene-based silanol polymer and copolymers. 4-Vinylphenyldimethylsilanol polymer and its styrene copolymers (PVPDMS) in solution were synthesized from corresponding precursor silane polymer or styrene copolymers by reacting with a dimethyldioxirane solution in acetone similar to the procedure described before (17-18). The precursor silane polymer and styrene copolymers were prepared by polymerization of 4-vinylphenyldimethylsilane monomer or copolymerization of the monomer with styrene in the presence of a free radical initiator (18). The characteristics and composition of styrene-based silanol polymers are listed in Table II.

Table II.   Styrene-Based Silanol Polymers

Code	Silanol Composition [a] (mol%)	$[\eta]$ [b] (dL/g)
PVPDMS-34	33.9	0.70
PVPDMS-60	60.0	0.76
PVPDMS-81	80.6	0.66
PVPDMS-100	100	0.55

[a] calculated from ^{1}HNMR spectra
[b] viscosities of precursor silane copolymers or homopolymer were measured in THF at 25 ± 0.01 °C.

Preparation of organic-inorganic polymeric hybrids. 2% (g/ml) solutions of PVPr, PVPy and PEOx in isopropanol were prepared. Blend solutions of PHMS with different organic polymers were prepared at weight ratios of 75/25, 50/50 and 25/75 by mixing appropriate amounts of each polymer in a common solvent while stirring. The resulting blend solutions were stirred at the room temperature overnight and then concentrated to approximately 2% concentration under reduced pressure at room temperature. Blend films were prepared by solution casting onto glass slides. After the solvent was slowly evaporated at room temperature, all the films were vacuum dried at 80 °C for 3 days unless otherwise specified in the text.

Preparation of blends and semi-interpenetrating polymer networks (semi-IPNs). 2 % (g/ml) solutions of styrene-based silanol copolymers and poly(N-vinylpyrrolidone) (PVPr) in chloroform were prepared. Blends were prepared by mixing appropriate amounts of each polymer solution while stirring. Mutual precipitation of two polymers took place while mixing. The precipitates were filtered and washed with chloroform and acetone, respectively, followed by vacuum drying at 80 °C to constant weight.

Thermal analysis. Differential scanning calorimetry (DSC) was performed by means of either the TA 2920 DSC or the Perkin-Elmer DSC-7 calorimeter. Sample weights of 8 ~ 12 mg and a heating rate of 20 °C/min were used. Glass transition temperature of blend was taken from the second scan unless otherwise specified in the text.

FT-IR spectroscopy. Fourier Transform infrared spectroscopy was performed with the use of the Perkin-Elmer 1600 series FT-IR or Digilab FTS-60 spectrometer. A minimum of 64 scans at a resolution of 2 cm^{-1} was signal-averaged. Samples for FT-IR studies were prepared by casting blend solutions onto KBr windows followed by vacuum drying at 80 °C for 3 days. For the precipitated inter-polymer complexes, KBr discs were prepared.

Results and Discussion

The conventional methods for organosilanol synthesis can be accomplished by the hydrolysis of corresponding silanes with various silicon functional groups in the presence of an acid or a base (8). This route, however, has some difficulty in application to the synthesis of silanol polymers where not only high conversion for polymer modification reaction is required but also the resistance of the silanol to self- or catalytic condensation during the preparation. A new convenient route for the synthesis of silanol polymers is developed by the direct oxidation of corresponding precursor polymers containing Si—H moiety (18,19). Polymer modification of the Si—H bond via reaction with a dimethyldioxirane solution in acetone proceeds rapidly and selectively. A series of novel inorganic siloxane- and styrene-based silanol polymers and copolymers were synthesized by the selective and rapid oxidation of precursor polymers and copolymers containing the Si—H bond with a dimethyldioxirane solution in acetone (Scheme I and II) (18,19). The silanol polymer or copolymers obtained in *situ* showed no tendency for condensation to siloxane in solution. The persistence of silanol polymers against the self-condensation in the solution was attributed to hydrogen bonded complexes between the silanols and the solvent. The silanol polymers obtained in solution were used for the preparation of organic-inorganic polymeric hybrids and blends by solution blending.

Scheme I.   Synthesis of Inorganic Siloxane-Based Silanol Polymer

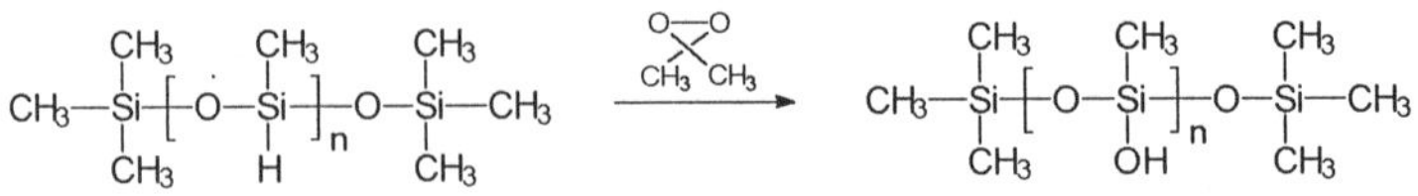

I   II

Scheme II.   Synthesis of Styrene-Based Silanol Polymer and Copolymers

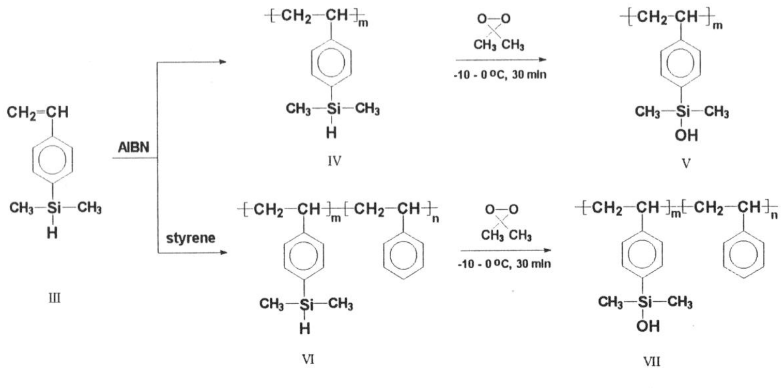

Organic-Inorganic Polymeric Hybrids

It is well known that siloxane polymers are essentially immiscible with almost all other polymers (20). Miscibility with organic polymers can be achieved by modification of siloxane polymer with strong hydrogen bond donor groups, such as 4-hydroxy-4,4-bis(trifluoromethyl) butyl to form strong inter-polymer hydrogen bonds (21) or by incorporating organic polymers within siloxane networks via sol-gel process (22-32) Since silanol functional groups may perform dual functions, either acting as a hydrogen bond donor to form inter-polymer hydrogen bonds with organic polymers to enhance miscibility or as reactive sites to condense with themselves to form siloxane crosslinkages. The relative strength of the inter-polymer and self-associated silanol hydrogen bonds as well as the condition for silanol self-condensation will be important for determining properties as well as the morphologies of hybrids. A preliminary investigation of 50/50 w/w blends of PHMS with PVPr or

PVPy organic polymer indicated that miscibility was accomplished in polymeric hybrids as evidenced by the clarity of blend films and the presence of a single glass transition temperature. Properties, such as T_g of the hybrid, depended mainly on the temperature at which the silanol crosslink takes place. For the hybrids vacuum dried at 40 °C for 3 days, a T_g at 54 °C for 50/50 PHMS/PVPr and 40 °C for 50/50 PHMS/PVPy were obtained. When the hybrids were dried in vacuum at 80 °C for 3 days, a higher T_g value was observed. Those results are shown in Table III. Infrared spectroscopy indicated that more self-associated silanol groups were transferred into siloxane crosslinks when the solid state crosslink was conducted at 80 °C. The great dependence of the hybrid properties on the temperature at which the silanol condensation takes place provides evidence for using caution in preparing these hybrids.

It is also important to mention that hydrogen bonds are thermally reversible. Therefore, the temperature at which the silanol self-condensation takes place should not only achieve maximum degree of crosslink density but also maintaining inter-polymer hydrogen bonds in the hybrid as intact as possible.

Table III. T_g **Values of the 50/50 w/w PHMS/PVPr and PHMS/PVPy Hybrids**

Blends	Ratio (w/w)	T_g (°C) drying at 40 °C	T_g (°C) drying at 80 °C
PHMS/PVPr	50/50	54	134
PHMS/PVPy	50/50	40	111

Table IV. **Blending Results of PHMS With a Variety of Organic Polymers**

Blends	Ratio (w/w)	Film clarity	T_g (°C)
PHMS/PVPr	75/25	Clear	170
PHMS/PVPr	50/50	Clear	167
PHMS/PVPr	25/75	Clear	164
PHMS/PEOx	75/25	Clear	72
PHMS/PEOx	50/50	Clear	61
PHMS/PEOx	25/75	Clear	45
PHMS/PVPy [a]	75/25	Clear	163
PHMS/PVPy [b]	75/25	Clear	160
PHMS/PVPy [a]	50/50	Clear	153
PHMS/PVPy [b]	50/50	Clear	154
PHMS/PVPy [a]	25/75	Clear	154
PHMS/PVPy [b]	25/75	Clear	155

[a] Films were cast from supernatant solutions.
[b] Inter-polymer complexes in the form of precipitates were obtained.

A variety of organic-inorganic polymeric hybrids were prepared and results are shown in Table IV (19). For hybrids vacuum dried at 80 °C, a single and constant T_g was obtained. Infrared spectroscopy indicated those strong inter-polymer hydrogen bonds between silanol and hydrogen bond acceptor groups of organic polymers were present and responsible for the miscibility in hybrids. Figure 1 shows the infrared spectra for hydrogen bonded silanol stretching vibration bands or carbonyl stretching vibration bands.

Styrene-Based Silanol Polymer Blends and Semi-IPNs

50/50 w/w Blends of styrene-based silanol polymer and copolymers were also prepared and results are shown in Table V (33). The strong inter-polymer hydrogen bonds resulted in the formation of inter-polymer complexes in the form of precipitates. The complexes had either a single T_g or no measurable T_gs. The glass transition temperature of the PVPDMS-34/PVPr complex (135 °C) is lower than those of calculated weight averaged value (146 °C). However, the glass transition temperature of PVPDMS-60/PVPr complex was found to be 183 °C — much higher than that of the calculated weight average value and also higher than that of higher T_g component polymer PVPr (T_g of PVPr is about 178 °C).

Table V.   Results of 50/50 w/w Blends of the PVPDMS/PVPr

Blend Code [a]	Solution Appearance	Complex Appearance	T_g (°C)	Calc. T_g (°C) [b]
PVPDMS-34	precipitation	clear	135	147
PVPDMS-60	precipitation	clear	183	150
PVPDMS-81	precipitation	clear	---- [c]	157
PVPDMS-100	precipitation	clear	---- [c]	163

[a] numeral following PVPDMS indicates mole % of silanol in the copolymers.
[b] calculated weight-average values.
[c] no T_g s were observed in the temperature range of 50 to 300 °C.

The PVPDMS-60, 81 and 100/PVPr inter-polymer complexes are insoluble in dimethylsulfoxide (DMSO) or dimethylformamide (DMF) after being isolated and dried at 80 °C for 4 days. However, the PVPDMS-34/PVPr complex is still soluble. The infrared spectroscopy study indicated that the stretching vibration band of Si-O-Si at 1044 cm^{-1} appeared in PVPDMS-60/PVPr, PVPDMS-81/PVPr and PVPDMS-100/PVPr, but was hardly detectable for the PVPDMS-34/PVPr complex (Figure 2). The appearance of the Si-O-Si absorption band suggested that the formation of siloxane cross linkages by condensation of silanol groups when the copolymers contained 60 mole% or more silanols. For PVPDMS-34/PVPr complex, the inter

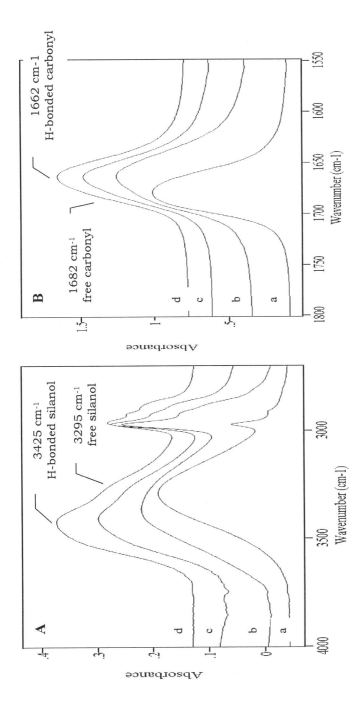

Figure 1. FT-IR spectra in the silanol (A) and carbonyl (B) stretching vibration regions for (a) PHMS and (b) 75/25, (c) 50/50 and (d) 25/75 PHMS/PVPr hybrids.

polymer hydrogen bond and relatively low concentration of silanol groups in the copolymer prevent the silanol condensation to form siloxane crosslinks.

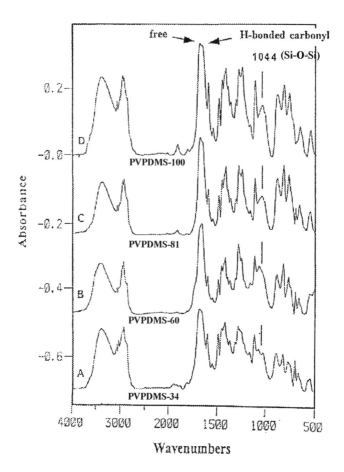

Figure 2. FT-IR spectra of (A) PVPDMS-34/PVPr complex, (B) PVPDMS-60/PVPr, (C) PVPDMS-81/PVPr and (D) PVPDMS-100/PVPr semi-IPNs.

The enhanced glass transition temperature of PVPDMS-60/PVPr could be attributed to the formation of siloxane crosslinks. The non measurable T_g's of the PVPDMS-81/PVPr and PVPDMS-100/PVPr may suggest that the obtained semi-IPNs have a higher crosslinking density consisting of contributions from both siloxane chemical crosslinkages and the physical crosslinkages of the strong inter-polymer

217

hydrogen bonds between the silanol and the amide carbonyl groups, resulting in a very small heat capacity change (ΔC_p) between the glass and rubbery states.

Conclusion

A new route for solid state crosslinking was developed by condensation of self-associated silanol groups in either binary miscible hydrogen-bonded polymer blends or organic-inorganic polymeric hybrids. The miscibility in either polymer blends or polymeric hybrids was achieved by the formation of strong inter-polymer hydrogen bonds between the silanol groups and the hydrogen bond acceptor groups of counter polymers. The solid state crosslinking was conducted by subsequent condensation of self-associated silanol groups in either miscible blends or polymeric hybrids without the presence of any external crosslinkers or catalysts. This new route for solid state crosslinking by utilizing the dual nature of the silanol functional groups led to the preparation of intimately mixed semi-interpenetrating polymer networks.

Acknowledgment: This research was supported in part by Grant No. DMR-9302375 from the National Science Foundation, Division of Materials Research, Polymer Group.

References

1 Sperling, L. H. *Interpenetrating Polymer Networks and Related Materials*, Plenum, New York, **1981**.
2 Klempner, D. and Berkowski, L. *Encyclopedia of Polymer Science and Engineering, Vol. 8*, 2nd Ed., Wiley, New York, **1987**.
3 Frisch, H. L., Klempner, D., Yoon, H. K. and Frisch, K. C., *Macromolecules*, **1980**, *13*, 1016.
4 Singh, S., Frisch, H. L. and Ghiradella, H., *Macromolecules*, **1990**, *23*, 375.
5 Kim, H. I., Pearce, E. M. and Kwei, T. K., *Macromolecules*, **1989**, *22*, 3374.
6 Chen, F. L., Pearce, E. M. and Kwei, T. K., *Polymer*, **1988**, *29*, 2285.
7 Yang, T. P., Kwei, T. K. and Pearce, E. M., *J. Appl. Polym. Sci.*, **1990**, *41*, 1327.
8 Noll, W. *Chemistry and Technology of Silicones*, Academic, New York, **1986**.
9 West, R., and Baney, R. H., *J. Inorg. Nucl. Chem.*, **1958**, 7, 297.
10 West, R., Baney, R. H. and Powell, D. L., *J. Am. Chem. Soc.*, **1960**, *82*, 6269.
11 Stone, F. G. A. and Seyferth, D., *J. Inorg. Nucl. Chem.*, **1955**, *1*, 112.
12 Reichstat, M. M., Mioc, U. B., Bogunovic, Lj. J. and Ribnikar, S. V., *J. Mol Struct.*, **1991**, *244*, 283.

13 Mioc, U. B., Bogunovic, Lj. J., Ribnikar, S. V. and Stankovic, N., *J. Serb. Chem. Soc.*, **1989**, *54*, 541.

14 Allred, L., Rochow, E. G. and Stone, F. G. A., *J. Inorg. Nucl. Chem.*, **1956**, *2*, 416

15 Harris, G. I., *J. Chem. Soc.*, **1963**, p. 5978.

16 Kerr, G. T. and Whitemore, F. C., *J. Am. Chem. Soc.*, **1946**, *68*, 2282.

17 **Lu**, S., Pearce, E. M., and Kwei, T. K., *J. Polym. Sci., Polym. Chem.*, **1994**, *32*, 2597.

18 Lu, S., Pearce, E. M. and Kwei, T. K. *Macromolecules*, **1993**, *26*, 3514.

19 Lu, S., Melo, M. M., Zhao, J., Pearce, E. M. and Kwei, T. K., *Macromolecules*, 1995, *28*, 4908.

20 Krause, S. in Polymer Blends; Paul D. R., Newman, S., Eds.; Academic Press, New York, **1978**, Vol. 1.

21 Chu, E. Y., Pearce, E. M., Kwei, T. K., Yeh, T. F., Okamoto, Y., Makromol. Chem., Rapid Comm., **1991**, *12*, 1.

22 Mark, J. E., Jiang, C. Y. and Tang, M. Y. *Macromolecules*, **1984**, *17*, 2613.

23 Tang, M. Y. Mark, J. E. *Macromolecules*, **1984**, *17*, 2616.

24 Clarson, S. J. and Mark, J. E. *Polym. Commun.* 1987, *28*, 249.

25 Brennan, A. B. and Wilkes, G. L. *Polymer*, **1991**, *32*, 733.

26 Huang, H. H., Orler, B. and Wilkes, G. L. *Macromolecules*, **1987**, *20*, 1322.

27 Wang, B., Wilkes, G. L. Hedrick, J. C. Liptak, S. C. and McGrath, J. E. *Macromolecules*, **1991**, *24*, 3449.

28 Saegusa, T., Chujo, Y. *Proceeding for the 33rd IUPAC Meeting on Macromolecules*, Montreal, **1990**.

29 Chujo, Y. and Saegusa, T. *Adv. Polym. Sci.*, **1992**, *100*, 11.

30 Landry, C. J. T., Coltrain, B. K., Brady, B. K. *Polymer*, **1992**, *33*, 1486.

31 Landry, C. J. T., Coltrain, B. K., Wesson, J. A., Zumbulyadis, N., Lippert, J. L. *Polymer*, **1992**, *33*, 1496

32 Fitzgerald, J. J. Landry, C. J. T. and Pochan, J. M. *Macromolecules*, **1992**, *25*, 3715.

33 Lu, S., Pearce, E. M. and Kwei, T. K. *Polymer*, **1995**, *36*, 2435.

BULK PHOTOPOLYMERIZATION

Chapter 14

Modeling and Experimental Investigation of Light Intensity and Initiator Effects on Solvent-Free Photopolymerizations

Michael D. Goodner and Christopher N. Bowman[1]

Department of Chemical Engineering, Campus Box 424, University of Colorado, Boulder, CO 80309–0424

The free radical photopolymerization of 2-hydroxyethyl methacrylate (HEMA) initiated by 2,2-dimethoxy-2-phenylacetophenone (DMPA) is studied using kinetic model predictions and experiments. Both homogeneous and spatially varying kinetic models are developed from species balances on reacting species in the polymerization system. The focus of this investigation is the effect of incident light intensity and photoinitiator concentration on the polymerization kinetics. In optically thin samples, the polymerization rate profiles are shown to have the classical square root dependence on both light intensity and initiator concentration in the absence of primary radical termination. Homogeneous model predictions incorporating primary radical termination show reduction and, at higher primary radical concentrations, elimination of autoacceleration and a reduction in the final conversion. These features can be attributed to the quenching effect that the primary radicals have on the macroradical population. The effect of sample thickness is investigated using a kinetic model incorporating spatial variations in one dimension. Optically thick samples show marked decreases in rate and conversion at the bottom of thick films, while use of photobleaching initiators can partially mitigate these effects.

Bulk polymerization of vinyl monomers provides a convenient route for production of polymeric materials used in a wide variety of applications. Materials formed via free-radical polymerizations include lithographic plates, photoresists, protective coatings, adhesives, and biomedical devices, such as controlled-release matrices and dental restorations (1-4). Because the monomers used in these processes are liquids of intermediate viscosity, no solvent is needed to carry out the polymerization in the liquid phase. In addition to any environmental benefits, the lack of a solvent reduces complications resulting from spurious and unwanted chain transfer reactions, along with providing greater rates of polymerization at low conversions.

Unfortunately, the kinetics of solvent-free polymerizations are greatly complicated at higher conversions in both the formation of linear polymers and the production of crosslinked systems. As viscosity and vitrification effects increase, autoacceleration begins, greatly increasing the rate of polymerization (1,3). For these highly exothermic polymerization reactions, increased rate translates into increased heat production. Without a solvent to dilute the system and remove the excess heat of

[1]Corresponding author: telephone: 303–492–3247; fax: 303–492–4341; e-mail: bowman@colorado.edu.

polymerization, charring and volatilization can potentially ruin the desired product. At even greater conversions, vitrification leads to autodeceleration and incomplete functional group conversion despite continued initiator presence and exposure to light. Unincorporated monomer can leach from the polymer system, leading to toxicity effects, especially in biomedical polymers. In crosslinked systems, unreacted pendent double bonds can react in the post-cure period, leading to brittleness and decreased permeability.

One method used to minimize viscosity and vitrification effects is to carry out the polymerizations at high initiation and polymerization rates (5). Due to the incorporation of many monomer units into each polymer molecule, volume shrinkage occurs upon polymerization. At high polymerization rates, conversion of monomer to polymer occurs at a much faster rate than the volume shrinkage. As a result, reacting molecules are afforded extra mobility, and the polymerizing system can reach much greater conversions than could be achieved while maintaining equilibrium volume shrinkage (1,5-7).

To reach higher polymerization rates, the population of growing polymer chains must be increased throughout the polymerization. The easiest route to achieve this goal is to increase the rate of initiation. However, in photopolymerizations, several effects pose difficulties when trying to reach higher initiation rates. Increasing the photoinitiator concentration in a monomer solution has the undesirable consequence of reducing light penetration. In a protective coating, this effect can result in a hard, tack-free surface that will not adhere to a delicate substrate because of the reduced polymerization at the bottom of the film. Increased photoinitiator content can also result in pronounced yellowing of the product that is associated with excess photoinitiator not incorporated into the polymer network. Increasing light intensity to reach higher initiation rates also has negative effects. In addition to the need for stronger, more expensive light sources, higher light intensities increase the population of primary radicals (radicals derived directly from photocleavage of the initiator). Large primary radical populations can lead to a phenomenon known as primary radical termination, which can reduce the polymerization rate.

To control photopolymerizations properly, the aforementioned negative effects must be well understood. This work first develops a homogeneous kinetic model, including diffusion-controlled kinetics, that describes the polymerization behavior in an optically thin sample. For this study, the model system is the polymerization of 2-hydroxyethyl methacrylate (HEMA) initiated by 2,2-dimethoxy-2-phenylacetophenone (DMPA). The influence of incident light intensity and photoinitiator concentration are briefly examined. These model predictions are compared to experimental rate profiles generated here for the HEMA/DMPA system and previously published for similar methacrylate systems (8). At higher light intensities, the effect of primary radical termination can be examined through the use of the homogeneous model. It is seen that as primary radical termination increases, autoacceleration is decreased and then eliminated, and the classical square root dependence of polymerization rate on light intensity no longer holds. To relate these effects to commercial polymerizations, the homogeneous model is incorporated into a one dimensional, spatially varying kinetic model to account for decreases in light intensity as a function of depth. 1D model predictions are presented for both conventional initiators (such as DMPA) and photobleaching initiators.

Homogeneous Model Development

In order to model photopolymerizations properly, the basic reaction mechanisms must be known. For the HEMA/DMPA system, neglecting chain transfer and inhibition/retardation, the overall mechanism is:

Initiation:

$$I \xrightarrow{h\nu} 2R\bullet \qquad (1a)$$

$$R\bullet + M \xrightarrow{k_i} P_1\bullet$$
(1b)

Propagation:

$$P_n\bullet + M \xrightarrow{k_p} P_{n+1}\bullet \qquad (1c)$$

Termination:

$$P_n\bullet + P_m\bullet \xrightarrow{k_t} P_{n+m} \quad (P_n + P_m)$$
(1d)

$$P_n\bullet + R\bullet \xrightarrow{k_{tp}} P_n R \qquad (1e)$$

In these equations, I represents the initiator species, M is monomer, P_n, P_m, and P_{n+m} are dead polymer chains, and $P_n R$ is dead polymer produced via primary radical termination. $R\bullet$ and $P_n\bullet$ represent primary radicals and growing macroradical chains of n repeating units, respectively.

Initiation is broken into two steps: photocleavage of initiator molecules to produce primary radicals (equation 1a) and the initiation of growing polymer chains (equation 1b). k_i is the kinetic constant for the chain initiation step, and the rate of step 1a is given by:

$$R_i = 2\phi\varepsilon I_0 b[I] \qquad (2)$$

where ϕ is the initiator efficiency, ε is the molar absorptivity of the initiator, $I_0 b$ is the incident light intensity in moles of photons per square centimeter per second and $[I]$ is the instantaneous initiator concentration. The 2 represents the generation of two primary radicals for each initiator molecule; in this study the two primary radicals are assumed to have identical diffusive and reactive characteristics. The properties for DMPA used in the model are listed in Table I, where the value for the efficiency, 0.6, is a typical value for free radical initiators.

Table I. Material and Kinetic Properties for DMPA and HEMA

Initiator Properties for DMPA:		
ε = 150 L/mol·cm	ϕ = 0.6	
Material Properties for HEMA:		
ρ_m = 1.073 g/cm^3	ρ_p = 1.15 g/cm^3	$[M]_0$ = 8.2 mol/L
T_{gm} = -60 °C	T_{gp} = 55 °C	
α_m = 0.0005 °C^{-1}	α_p = 0.000075 °C^{-1}	
Kinetic Parameters for HEMA:		
R = 4 L/mol^a		
k_{p0} = 1000 L/mol·s	A_p = 0.66	f_{cp} = 0.042
k_{t0} = 1.1 x 10^6 L/mol·s	A_t = 1.2	f_{ct} = 0.060
k_{i0} = 1000 L/mol·s	A_i = 0.66	f_{ci} = 0.042
k_{tp0} = varies	A_{tp} = 0.66	f_{ctp} = 0.042

^{a}The reaction-diffusion parameter, R, is from C. N. Bowman (unpublished data).

In the propagation step (equation 1c), monomer is added to the growing polymer chains with kinetic constant k_p. All growing polymer chains ($n \geq 1$) are assumed to be identical, *i.e.*, they have the same reactivities. Termination can occur through two mechanisms. Bimolecular termination (equation 1d) can occur through either combination or disproportionation. Since the mode of bimolecular termination does not influence the kinetics, a single termination kinetic constant (k_t) accounting for both modes will be used. In primary radical termination (equation 1e), a growing polymer chain is terminated by reacting with a primary radical produced from initiator photocleavage. The primary radical termination kinetic constant (k_{tp}) will in general be different from k_t, due to the difference in reactivity and diffusivity between macroradicals and primary radicals.

Once the general mechanism is known, the model development is straightforward. Species balances are performed for the six reacting species in the system: initiator, primary radicals, monomer, growing macroradicals, dead polymer, and primary radical-terminated polymer. An example of such a species balance for the growing macroradicals is:

$$\frac{d[P_n \bullet]}{dt} = k_i [R \bullet][M] - 2k_t [P_n \bullet]^2 - k_{tp}[R \bullet][P_n \bullet] \quad (3)$$

The six species balances, coupled with kinetic constant expressions and correlations between conversion and fractional free volume, are integrated numerically to provided concentration profiles. From these data, rate-conversion-time relations can be developed.

The expressions for the kinetic constants used in this work are those derived for diffusion-controlled kinetics (*9-11*). Following the work of Anseth and Bowman (*12*) the expressions for the propagation and termination kinetic constants are:

$$k_p = \frac{k_{p0}}{\left(1 + e^{A_p\left(1/f - 1/f_{cp}\right)}\right)} \quad (4a)$$

$$k_t = \frac{k_{t0}}{\left(1 + \left(\frac{Rk_p[M]}{k_{t0}} + e^{-A_t\left(1/f - 1/f_{ct}\right)}\right)^{-1}\right)}$$

(4b)

In the expression for k_p, k_{p0} is the true kinetic constant for propagation in the absence of all diffusional limitations, A_p is a parameter governing the rate of decrease of k_p in the diffusion-controlled regime, f is the fractional free volume of the polymerizing system and f_{cp} is the critical fractional free volume for propagation, *i.e.*, the fractional free volume value at which k_p is half k_{p0}. The forms for k_i and k_{tp} are assumed to be identical to equation 4a, with similar parameters governing their diffusional dependence. The expression for k_t has an extra term to account for reaction diffusion. In that term, R is the reaction diffusion parameter and [M] is the instantaneous monomer concentration (or, more correctly, unreacted double bond concentration, in the case of multifunctional monomers).

A method for determining the kinetic parameters for propagation and bimolecular termination has previously been presented (*13*). The technique requires a single experimental rate versus conversion profile and will not be reiterated here. The values determined for HEMA in that work will be used for this work and are found in Table I. Finding values for the parameters governing chain initiation and primary radical termination is not as straightforward. In this work, two assumptions will be made. First, since chain initiation and primary radical termination both rely on the diffusion of small molecules (R• and M for chain initiation and R• for primary radical termination), their diffusional characteristics will be assumed to be identical to propagation. Thus $A_{tp} = A_i = A_p$ and $f_{ctp} = f_{ci} = f_{cp}$. Second, k_{i0} will be approximated by k_{p0}, since it is essentially a radical propagation reaction. k_{tp0} will not be fixed. In fact, its value will be varied in the simulations to determine the relative effect of primary radical termination on the polymerization rate profile.

Experimental

For the experiments, HEMA was obtained from Aldrich (Milwaukee, WI) and was dehibited prior to use. DMPA was obtained from Ciba-Geigy (Hawthorne, NY) and used as received. The polymerizations were performed in a Perkin-Elmer differential scanning calorimeter (DSC) equipped with a photoaccessory. The photoaccessory includes a monochromator, and 365 nm ultraviolet light was used to initiate the polymerizations. A light intensity of 3.8 mW/cm^2 was used for the polymerizations. The polymerization rate is determined by monitoring the rate of heat evolution and normalizing by the standard heat of reaction for methacrylates, -13.1 kcal/mol (*14*). Samples were limited to 2 to 3 milligrams to ensure the validity of the thin film approximation for light absorption. A nitrogen purge was used prior to and during the DSC runs to prevent oxygen inhibition of the free radical polymerizations.

Results and Discussion

Homogeneous Kinetic Model. The validity of the homogeneous model for predicting rate profiles at low light intensities and a single low initiator concentration has already been established (*15*). When the initial initiator concentration is increased, the polymerization rate increases accordingly, as shown in Figure 1. The simulation predicts the rate profiles rather well, giving good agreement in the absence of diffusional limitations, throughout the autoacceleration and autodeceleration regimes, up to the final limiting conversion. The simulation shows the classical square root dependence of polymerization rate on initiation rate in the absence of primary radical termination. As the light intensity is held constant, the rate of initiation is proportional to the initiator concentration (equation 2). Thus, the rate profiles scale with the square root of the initiator concentration. This scaling effect has also been previously reported for HEMA polymerizations initiated thermally by 2,2'-azobis(isobutyronitrile) (*8*).

The rate of initiation can also be changed by varying the incident light intensity while the initiator concentration is held constant. Homogeneous model predictions for a three order of magnitude range of light intensities at a fixed initiator concentration are shown in Figure 2. Once again, the square root dependence of polymerization rate on initiation rate is seen. The reason for this dependence becomes clear when the species balance for the macroradical concentration (equation 3) is examined. After the initial macroradical population is established, the pseudo-steady state assumption can be applied to equation 3. In the absence of primary radical termination, chain initiation must balance bimolecular termination. Equating the first and second terms on the right hand side of equation 3, it is seen that the macroradical concentration varies with the square root of the primary radical concentration, and thus, initiation rate.

When primary radical termination becomes the dominant mechanism of termination, the square root dependence is no longer seen, as witnessed in Figure 3. At

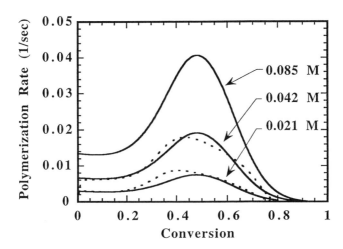

Figure 1. Experimental polymerization rate profiles (dotted lines) compared to model predictions (solid lines) for several different initial initiator concentrations. Primary radical termination effects are not included in the simulation ($k_{tp0} = 0$).

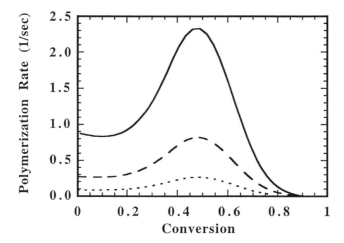

Figure 2. Polymerization rate as a function of incident light intensity as predicted by the homogeneous kinetic model. The lines correspond to 100 mW/cm^2 (solid), 10 mW/cm^2 (dashed) and 1 mW/cm^2 (dotted). The initiator is 0.021 M (0.5 weight percent) DMPA; primary radical termination is not included ($k_{tp0} = 0$).

lower light intensities (the 1 mW/cm^2 curve), the balance between chain initiation and bimolecular termination still dictates the macroradical concentration, and the square root dependence still holds. The rate curve resembles the profiles seen previously in Figure 2. As the light intensity is increased to 10 mW/cm^2, primary radical termination effects become apparent. At lower conversions, the rate still scales with light intensity. Once autoacceleration begins, however, the correlation breaks down. The scaled polymerization rate is below the value predicted by the pseudo-steady state analysis of equation 3. As the light intensity is increased, the reduction in rate becomes greater and manifests earlier. At the highest light intensity (1000 mW/cm^2), the rate is greatly depressed, to the point of autoacceleration being nearly eliminated.

The reason behind this phenomenon is as follows: at low light intensities, the primary radical population is relatively low. In equation 3, bimolecular termination (the second term) dominates primary radical termination (the third term). The balance determining the macroradical population is still between chain initiation and bimolecular termination. At high light intensities (high primary radical concentrations), primary radical termination is the dominant mechanism of termination and balances chain initiation. Setting the first and third terms equal on the right hand side of equation 3 causes the primary radical concentrations to cancel, and the macroradical concentration becomes independent of primary radical concentration and therefore light intensity. Not only is the macroradical concentration now independent of light intensity, but it is also reduced due to the quenching effect of the primary radicals.

At intermediate light intensities (such as 10 or 100 mW/cm^2), bimolecular termination dominates at lower conversions, but as k_t starts to decrease in the autoacceleration region, bimolecular termination drops while primary radical termination remains constant. Thus, the balance shifts from bimolecular to primary radical termination. Primary radicals now quench the build-up of macroradicals, leading to large reductions in the height of the autoacceleration peak. Further, more detailed studies on the effects of primary radical termination on polymerization rate are reported in Goodner and Bowman (*16*).

Kinetic Model with Spatial Variations. While the previous examinations give insight into the polymerization behavior response as light intensity and initiator concentration are changed, they reflect the polymerization in an optically thin film. In real processes, however, the light intensity can decrease markedly with depth into a film, especially at low light intensities and high initiator or pigment concentrations (*17*). In order to treat the polymerization kinetics properly in these optically thick films, a slightly different approach must be taken. The simplest route is to divide the thick sample into many optically thin slices, over which the light intensity does not change appreciably due to absorbance by chromophores such as initiator molecules. The homogeneous kinetic model can then be applied to each slice at a given time step, and the light intensity profile across all slices can be recomputed between time steps (by applying Beer's law). This scheme is reflected in Figure 4; the resulting model incorporating spatial variations will hereafter be referred to as the '1D Model'. For the preliminary results of the 1D model reported here, the simplifying assumptions of no mass and heat transfer across the slices are made. Additionally, the sample is assumed to be isothermal, *i.e.*, all slices are at the same temperature and the temperature does not change as the reaction proceeds. While these assumptions will be invalid for either high polymerization rates or high double bond conversions, they provide a good first approximation and give insight into the general trends in the polymerization behavior.

The rate profile predicted by the 1D model for low light intensity and initiator concentration is shown in Figure 5. It is apparent that the thin-film approximation for light intensity still holds across a 3 mm thick film for these low values. The rate profiles at the top and bottom of the film are nearly identical. A slight delay - approximately 20 seconds - is seen for both the onset of autoacceleration and the time for maximum rate (which occurs around 170 seconds at the top of the sample); this

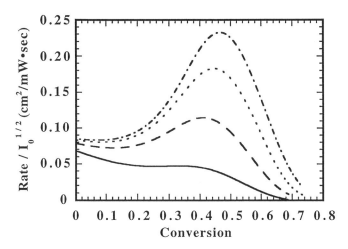

Figure 3. Scaled polymerization rate as a function of incident light intensity as predicted by the homogeneous kinetic model. The lines correspond to 1000 mW/cm^2 (solid), 100 mW/cm^2 (dashed), 10 mW/cm^2 (dotted) and 1 mW/cm^2 (dash-dot). The initiator is 0.021 M (0.5 weight percent) DMPA; primary radical termination is included (k_{tp0} = 1.0 x 10^7).

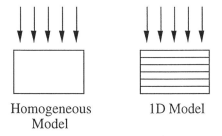

Homogeneous   1D Model
Model

Figure 4. Comparison of the homogeneous kinetic model to the 1D model. Arrows indicate the direction of light illumination.

feature is caused by the slightly reduced light intensity at the bottom of the film. From a processing point of view, though, the variations are slight enough to treat the sample as uniform, and reaching the specified conversion at the top of the film would be sufficient to satisfy product specifications.

If the initiator concentration is increased to 0.05 M (corresponding to a little over 1 weight percent), the heterogeneity seen throughout the sample is quite pronounced. Figure 6a shows the rate profile with respect to time and depth into the sample. At the top of the film, a normal profile is seen, with no diffusion-controlled kinetics up to 20 seconds, followed by autoacceleration and autodeceleration down to a small, near zero, rate. The polymerization behavior at the bottom of the film shows a similar profile, but with a marked decrease in rate and greatly delayed onset of autoacceleration. The maximum rate is not reached at the bottom of the film until 100 seconds, and the maximum rate that is achieved is less than a quarter of the value seen at the top of the film.

Another way to examine the influence of optical thickness resides in the conversion versus time and depth data shown in Figure 6b. After polymerizing for 75 seconds, 75 percent of the monomer in the system is incorporated into polymer chains at the top of the film. In contrast, the bottom of the film shows less than 30 percent conversion, leaving a great deal of monomer unincorporated and leachable, devastating the mechanical properties of the film and raising serious toxicity concerns. Remedying this problem requires considerable effort. For example, if a conversion of 60 percent is needed to reach product specifications, this condition is met after 50 seconds at the top of the film. However, because of the attenuated light intensity, sixty percent conversion is not reached until after 150 seconds at three millimeters depth, causing a three-fold increase in processing time.

Increasing the initiator content further would merely serve to increase the optical density of the sample, limiting light penetration and exaggerating the problems seen in the previous figures. If higher initiator concentrations are needed, one possible solution to the problem is the use of a photobleaching initiator. Upon cleavage, a photobleaching initiator loses some or all of its absorptivity at the illuminating wavelength.

Model predictions using 0.5 M photobleaching initiator are shown in Figure 7. In this set of data, the initiator has the same efficiency and molar absorptivity at 365 nm, but is assumed to lose all absorptivity at 365 nm after photocleavage. Initially, a high rate of polymerization is reached in a small region near the top of the film into which the light can effectively penetrate. After polymerization in this region, the polymerization proceeds along a polymerization front. The front propagates into the film as the excess initiator in the polymerized region cleaves and becomes transparent. While the polymerization proceeds at relatively slow rates compared to the lower initiator concentrations where the thin-film approximation holds, the use of a photobleaching initiator allows higher concentrations to be used to give higher initiation rates.

Conclusions

In this paper, the effect of incident light intensity and photoinitiator concentration on the radical photopolymerization of HEMA initiated by DMPA has been studied using both homogeneous and spatially varying kinetic models. These models are based on numerically integrating species balances of the reacting species in the system. Additionally, experimental studies have been performed to verify the accuracy of the homogeneous kinetic model predictions.

From the homogeneous model predictions, the effect of primary radical termination can be seen (as shown in Figure 3). At high light intensities and high initiator concentrations, primary radical termination reduces the increase in polymerization rate afforded by autoacceleration and decreases the final conversion.

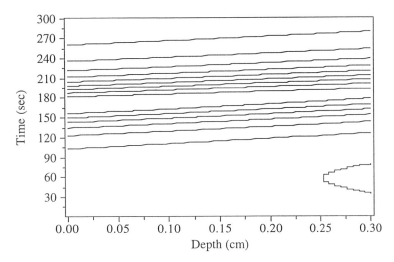

Figure 5. Rate versus time and depth as predicted by the 1D model for the polymerization of HEMA in a 3 mm thick film. Contours are spaced every 0.004 sec^{-1}. $[I]_0 = 0.005$ M DMPA and $I_0 = 4.0$ mW/cm^2.

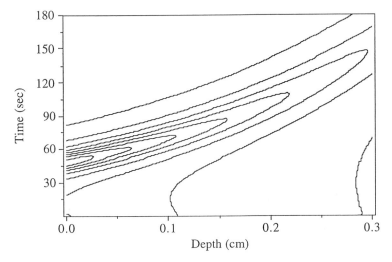

Figure 6a. Rate versus time and depth as predicted by the 1D model for the polymerization of HEMA in a 3 mm thick film. Contours are spaced every 0.02 sec^{-1}. $[I]_0 = 0.05$ M DMPA and $I_0 = 4.0$ mW/cm^2.

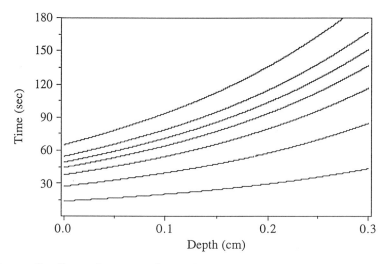

Figure 6b. Conversion versus time and depth for the polymerization shown in Figure 6a. Contours are spaced every 10 percent conversion.

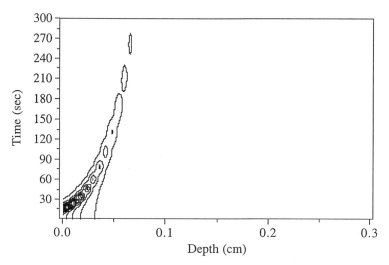

Figure 7. Rate versus time and depth as predicted by the 1D model for the polymerization of HEMA in a 3 mm thick film. Contours are spaced every 0.04 sec^{-1}. $[I]_0 = 0.5$ M photobleaching initiator and $I_0 = 4.0$ mW/cm^2. The beaded appearance along the polymerization front is a phenomenon of interpolation to the grid by the presentation software; it does not represent any physical phenomenon.

The effects of sample thickness are shown by the 1D model results. As initiator concentration is increased, light absorption throughout the film decreases the light intensity and, hence, the rate of polymerization at the bottom of the film. This decreased rate results in lower conversions and increased processing times, which must be accounted for in the production of commercially viable products. While not completely eliminating the problem, the use of photobleaching initiators mitigates the effects of high optical densities.

Acknowledgments

The authors would like to acknowledge the National Science Foundation for its support of this work through the Presidential Faculty Fellowship to CNB (CTS-9453369) and a graduate fellowship to MDG as well as 3M and the Camille Dreyfus Teacher-Scholar program.

Literature Cited

1. Kloosterboer, J. G. *Adv. Polym. Sci.* **1988**, *84*, 1-61.
2. Kannurpatti, A. R.; Peiffer, R. W.; Guymon, C. A.; Bowman, C. N. in *Critical Reviews of Optical Science and Technology: Polymers in Optics: Physics, Chemistry, and Applications*; Lessard, R. A. and Frank, W. F., Ed., **1996**; Vol. CR63, pp 136.
3. Decker, C. *Prog. Polym. Sci.* **1996**, *21*, 593-650.
4. Anseth, K. S.; Newman, S. M.; Bowman, C. N. *Adv. Polym. Sci.* **1995**, *122*, 177-217.
5. Bowman, C. N.; Peppas, N. A. *Macromolecules* **1991**, *24*, 1914-1920.
6. Anseth, K. S.; Bowman, C. N.; Peppas, N. A. *J. Polym. Sci. Polym. Chem.* **1994**, *32*, 139-147.
7. Kloosterboer, J. G.; Lijten, G. F. C. M. in *Cross-Linked Polymers: Chemistry, Properties, and Applications*; Dickie, R. A., Labana, S. S. and Bauer, R. S., Ed.; American Chemical Society: Washington, D.C., **1988**; Vol. 367, pp 409-426.
8. Scranton, A. B.; Bowman, C. N.; Klier, J.; Peppas, N. A. *Polymer* **1992**, *33*, 1683-1689.
9. Marten, F. L.; Hamielec, A. E. *J. Appl. Polym. Sci.* **1982**, *27*, 489-505.
10. Marten, F.; Hamielec, A. in *Polymerization Reactors and Processes*; Henderson, J. and Bouton, T., Ed.; American Chemical Society: Washington D. C., **1978**; Vol. 104, pp 43-69.
11. Soh, S. K.; Sundberg, D. C. *J. Polym. Sci. Polym. Chem.* **1982**, *20*, 1299-1313.
12. Anseth, K. S.; Bowman, C. N. *Polym. React. Eng.* **1993**, *1*, 499-520.
13. Goodner, M. D.; Lee, H. R.; Bowman, C. N. *Ind. Eng. Chem. Res.* **1997**, *36*, 1247-1252.
14. Cook, W. D. *Polymer* **1992**, *33*, 2152-2161.
15. Kannurpatti, A. R.; Goodner, M. D.; Lee, H. R.; Bowman, C. N. in ; Scranton, A. B., Ed.; American Chemical Society: Washington, D.C., **1997**.
16. Goodner, M. D.; Bowman, C. N. *Polym. React. Eng.* **in preparation**.
17. Cook, W. D. *J. Appl. Polym. Sci.* **1991**, *42*, 2209-2222.

Chapter 15

Gas Phase and Cluster Studies of the Early Stages of Cationic Polymerization and the Reactions with Metal Cations

Y. B. Pithawalla and M. Samy El-Shall[1],

Department of Chemistry, Virginia Commonwealth University, Richmond, VA 23284–2006

In this chapter, we present several examples of gas phase and intracluster cationic polymerization studies. Intracluster polymerization refers to a process where neutral clusters of selected monomer molecules are formed in the gas phase and the polymerization reactions are initiated by chemical or physical methods. In the gas phase at low pressures, eliminative polymerization may predominate. However, in the same systems at high pressure, the ionic intermediates may be stabilized and addition without elimination may occur. In clusters, evidence for both stabilization by fast evaporation (boiling off) of solvent molecules and for eliminative polymerization has been established.

The ion-molecule reactions of isobutene have been studied both in the gas phase and within isobutene clusters. Two parallel reactions producing the t-butyl carbocation ($C_4H_9^+$) and a covalent radical cation dimer ($C_8H_{16}^{+\cdot}$) are observed. Trends in the kinetics and thermochemistry as a function of reaction steps are reported.

In the metal-catalyzed systems, reactions of Ti^+, Zr^+ and Mo^+ with isobutene have been studied. Implications of these studies to bulk and solution polymerization are discussed. The gas phase and cluster approaches are important not only for a fundamental understanding of the early stages of polymerization, but also for the development of new materials with unique properties such as defect-free polymeric films of controlled composition.

I. Introduction

Understanding the fundamental mechanisms that govern polymerization reactions is an important requirement for the development of a number of scientific disciplines involving chemical reactivity and reaction mechanisms as well as practical applications including

[1]Corresponding author.

industrial processes, radiation chemistry, interstellar synthesis, and the development of new materials that possess novel properties.[1-3]

In order to arrive at a fundamental understanding of the processes involved in a typical polymerization system, it is important to understand the early stages of polymerization reactions in the gas phase and proceed by studying these reactions in *clusters* in order to integrate the new information with the existing knowledge from condensed phase. Clusters represent an intermediate regime between the gas and condensed phases and therefore, provide excellent opportunities to observe size-specific reactions and examine the dependence of chemical reactivity and solvent effects on the state of aggregation in the clusters.[4,5]

Intracluster polymerization refers to a process where neutral clusters of selected monomer molecules are formed in the gas phase and the polymerization reactions are initiated by chemical or physical methods.[6-20] The neutral clusters are synthesized using supersonic cluster beam techniques.[4-10] Cationic polymerization can be initiated within the clusters following their ionization either by electron impact (EI) or laser multiphoton ionization (MPI). Other initiation mechanisms may involve metal-catalyzed reactions where metal ions, atoms and clusters are generated by laser vaporization techniques and, then allowed to react with the monomer clusters. Ion-molecule addition and elimination reactions take place within the cluster ions, resulting in a product ion distribution that reflects both the stability of the polymeric ions and the kinetics of the reaction. Photodissociation experiments on mass-selected ions can characterize the structures of the polymeric ions, and hence, measure the extent of the propagation reaction. The processes involved in intracluster polymerization are shown schematically in Figure 1.

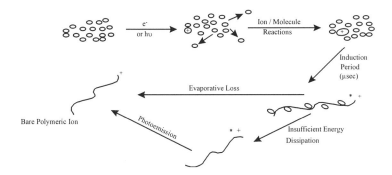

Figure 1. Typical processes involved in intracluster polymerization

Cationic polymerization in clusters possesses a distinct form of chemistry that has not been explored in much detail. In the gas phase at low pressures, eliminative polymerization may predominate.[20-24] However, in the same systems at high pressure, the ionic intermediates may be stabilized and addition without elimination may occur.[20-29] In clusters, evidence for both stabilization by fast evaporation (boiling off) of solvent molecules and for eliminative polymerization has been established. These results showed that sequential polymerization reactions, with 2-10 condensation steps, can occur on a time scale of a few microseconds following the ionization of the neutral clusters.[6-20] This occurs despite the fact that condensation reactions are exothermic, and the energy released could dissociate the cluster. The competition between the condensation reactions and monomer evaporation can control the ultimate size that the polymer can reach in the cluster. The evaporation process is analogous to the evaporative control of temperature by adding volatile solvents (internal coolant) to ionic polymerization mixtures which is a well-established method in polymer chemistry.[1-2]

Intracluster ionic polymerization permits the effects of various factors which influence the ultimate size the polymer can reach in a cluster to be studied in a controlled manner. These include the study of the initiation and propagation mechanisms, competitive channels, the effects of inhibitors, and termination. The conversion of van der Waals clusters into size-specific covalent-bonded polymeric species has also some important implications for the development of new materials with unique properties. For example, the polymeric species could be deposited from cluster beams in a size-selected manner on metal or semiconductor surfaces. Systematic experimentation on a range of important polymers and matched metal substrates would make available for the first time a base of results upon which models for polymer/surface interactions and the properties of future polymeric materials could be reliably assessed.

In this paper, we describe several examples of gas phase and intracluster polymerizations with special emphasis on the reactions with metal cations. First, we consider the reactions of the radical cations within uncatalyzed systems. Specifically, we discuss the gas phase ion chemistry of isobutene. Second, we discuss the reactions induced by metal cations and the potential for using *in situ* gas phase catalysts for intracluster polymerization. Finally, we discuss the implications of the gas phase and cluster studies to the condensed phase systems.

II. Experimental

The gas phase reactions have been studied by the Laser Vaporization High Pressure Mass Spectrometry (LV-HPMS) technique.[15,27] In this method, the gas phase reactions of the selected monomer are probed by mass spectrometry following the laser vaporization / ionization of a metal target placed inside a high pressure cell (1-2 torr) as described in recent publications.[15,27] A schematic diagram of the experimental set up is shown in Figure 2-a. Briefly, the metal cations are generated by focusing the output of an excimer laser (XeCl, 308 nm) or the second harmonic of a Nd:YAG laser (532 nm, energy < 10 mJ/pulse) pulsed at 5 Hz on a selected metal rod placed inside the HPMS source. Isobutene vapor or isobutene / Ar mixture of known composition is admitted to the HPMS source at selected pressures via an adjustable needle valve. Ions exit the source through a 0.02 cm diameter hole and are analyzed using a quadrupole mass filter (Extrel C-50) which is mounted coaxially to the ion exit hole. The operating pressure in the mass spectrometer region is typically $(2-8) \times 10^{-6}$ torr. For uncatalyzed reactions, the monomer ions are generated within the high pressure cell, containing a suitable carrier gas such as N_2, by using a pulsed electron gun.[28] or laser multiphoton ionization.[29]

For cluster polymerization studies, neutral clusters are produced using supersonic cluster beam techniques and the polymerization reactions are initiated either by EI ionization of the clusters or by the metal cations produced by pulsed laser vaporization of a metal target.[14,16,30] A schematic diagram of the experimental set up in case of the metal ions is shown in Figure 2-b. The essential elements of the apparatus are jet and beam chambers coupled to a coaxial quadrupole mass filter. During operation 2%-10% isobutene in He or Ar carrier gas at a pressure of 1-2 atm is expanded through a conical nozzle (0.5 mm diameter) in pulses of 150-250 μs duration at repetition rates of 5-10 Hz. The metal cations are generated by direct laser vaporization/ionization of the metal target using the output of the second harmonic of a Nd:YAG laser. The metal ions are mixed with the neutral isobutene clusters generated by the pulsed adiabatic expansion and the jet is skimmed by a 3 mm conical skimmer and passed into a high vacuum beam chamber maintained at 8×10^{-8} to 1×10^{-7} torr during operation. The cluster ion beam is then analyzed using a quadrupole mass filter. The amplified signal from the particle multiplier is processed using a boxcar integrator set to sample the arrival times appropriate for the detected ions. The pulsed nozzle valve, the laser and the boxcar are all synchronized through a series of delay generators.

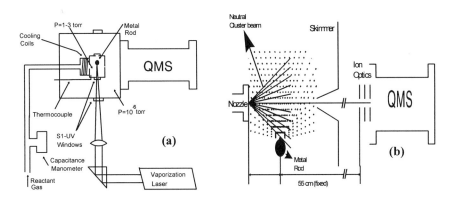

Figure 2. (a). Schematic for the Laser Vaporization-High Pressure Mass Spectrometry (LV-HPMS) source. (b). Schematic for the metal ion/clusters experiment.

III. Results and Discussion

A. Uncatalyzed Systems. Gas Phase Ion Chemistry of Isobutene. Isobutene is known to be polymerized in the bulk phase only by cationic mechanisms[1]. It has been demonstrated that this polymerization can take place in the gas phase.[1,27] Although it is generally accepted that the initiator of isobutene polymerization is the t-butyl carbocation ($C_4H_9^+$), the growing chain $C_4H_9^+(C_4H_8)_n$ has never been directly observed. Using HPMS, we characterized the energetic and kinetic properties of the key species in the gas phase isobutene system.[28] In this case, the molecular ions are formed by chemical ionization induced by a pulsed electron gun within a mixture of isobutene in N_2 at total pressures of 2-5 torr. The results indicate that the gas phase reactions in C_4H_8 after ionization follow two typical paths, radical ion chemistry (of $C_4H_8^{+\cdot}$ and its products), and carbocation chemistry (of t-$C_4H_9^+$ and its products). Figure 3 shows the ion profiles in a typical reaction mixture of C_4H_8 in N_2 following the ionization pulse in the HPMS experiment. The two parallel reactions (1 and 2) producing the $C_4H_9^+$ and the dimer $C_8H_{16}^{+\cdot}$ are clearly observed. We applied variable temperature HPMS to obtain basic information on both paths.[28]

$$C_4H_8^{+\cdot} + C_4H_8 \begin{cases} \longrightarrow C_4H_9^+ + C_4H_7^{\cdot} \quad (1) \\ \longrightarrow C_8H_{16}^{+\cdot} \quad (2) \end{cases}$$

Branching between the two products determines polymerization products and yields. We found that the branching ratio is strongly temperature dependent, with the radical dimer being dominant at 300 K, but the carbocation branch (leading to higher polymers) dominant above 400 K.

In the radical ion branch, the nature of the dimer product was not well established previously. Our results showed that the dimer is a covalent adduct with a dissociation energy >132 kJ/mol which is significantly larger than typical binding energies of non-covalent ion-molecule complexes. We used charge transfer bracketing and found the dimer to be the ion of a branched octene with an IP of 8.55±0.15 eV. A possible structure of the $C_8H_{16}^{+\cdot}$ dimer may be $(CH_3)_2C \cdot CH_2CH_2C^+(CH_3)_2$ which could be produced by tail-to-tail addition of $C_4H_8^{+\cdot}$ to C_4H_8 which should be favored sterically.

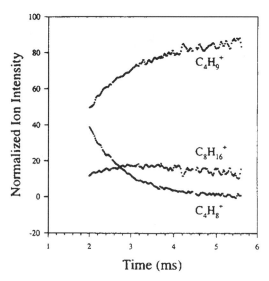

Figure 3. Normalized ion profiles in a reaction mixture of 4.575 torr of N_2 and 0.00006375 torr of isobutene at 501 K.

In the carbocation branch, we found that the addition reaction (3), for n=1, is reversible at 230 - 400 K, with ΔH^o = - 95.8 kJ/mol and ΔS^o = -136.8 J/mol K.

$$C_4H_9^+(C_4H_8)_{n-1} + C_4H_8 \longrightarrow C_4H_9^+(C_4H_8)_n \qquad (3)$$

Of particular interest is the second polymerization step, n=2 in reaction (3), that showed an unusually large thermochemical change, with ΔH^o = -101.3 kJ/mol and ΔS^o = -203.7 J/mol K. The large entropy change is characteristic to sterically hindered reactions.

In parallel to the thermochemistry, the kinetics of the second step also showed an anomaly. While the reaction efficiency of the first step was near unity, the second step was slower by orders of magnitude, with a reaction efficiency of 0.005 and with large negative temperature coefficient of $(T^{-12} - T^{-16})$, one of the largest reported in chemical kinetics. The third and fourth steps were still slower, but with smaller temperatures coefficients. The slower rate and large negative temperature coefficient of the second polymerization step are characteristic of reactions about sterically hindered centers, and result from entropy barriers due to the freezing of internal rotors in the transition state.

Ion Chemistry within Isobutene Clusters. Isobutene clusters were generated using supersonic beam expansion of isobutene in He or Ar (as a carrier gas), followed by electron impact ionization.[16] The mass spectrum following the 70 eV, EI - ionization is shown in Figure 4. The main sequence corresponds to the species $(C_4H_8)_n^{+\cdot}$ with n = 1 - 50. Another ion sequence corresponding to the doubly charged isobutene clusters $(C_4H_8)_n^{++}$ appears in

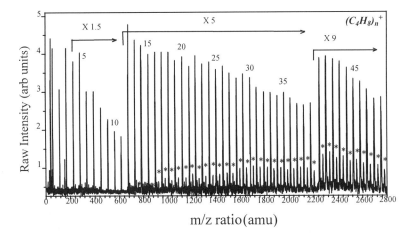

m/z ratio(amu)

Figure 4. 70-eV, EI mass spectrum of isobutene clusters $(C_4H_8)_n$ produced from isobutene/He expansion. Beam composition is 11% isobutene in He, P_0=24 psi, T_0=298 K. * indicates doubly charged clusters which are generated only for n ≥ 25.

the spectrum only past n ≥ 25. The identity of the doubly charged ions was verified by varying the electron energy of the EI - ionization. Figure 5 shows the mass spectrum obtained by 26 eV and 79 eV, EI - ionizations. As expected at lower electron energy (less than 27 eV) no doubly charged ions are generated due to the higher ionization potential of these species.

It is interesting to note that the isobutene clusters product under cold conditions do not show significant population of the $C_4H_9^+$ series i.e. $(C_4H_9)(C_4H_8)_n^+$. It appears that under cold cluster conditions and low energy ionization, no extensive evaporation of neutral or radical species takes place from the clusters. This suggests that reaction (1) occurs without complete removal of $C_4H_7^{\cdot}$ from the cluster. A similar argument can be made for the formation of isobutene dimer cations i.e. $C_8H_{16}^{+\cdot}$ within the cluster. In this case, no radical species are produced and the covalent dimer will have the same mass as the reactant ion and molecule. The resulting ions can therefore be considered as:

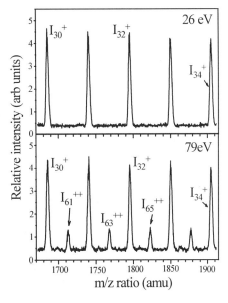

Figure 5. 26 eV and 79 eV EI - ionization energy mass spectra produced from isobutene/He expansion. Beam composition is 11% isobutene in He, P_0=24 psi, T_0=298 K.

$$- \overset{|}{\underset{|}{C}} (CH_2 - \overset{|}{\underset{|}{C}})_n^+ \cdots\cdots C_4H_7\cdot$$

and

$$\cdot \overset{|}{\underset{|}{C}} - CH_2 - CH_2 - \overset{|}{\underset{|}{C}} (CH_2 - \overset{|}{\underset{|}{C}})_n^+$$

These structures suggest that intracluster propagation takes place through the growing carbocation with the radical species ($C_4H_7\cdot$) still weakly attached to the cluster. Similarly, the generation of the distonic isobutene dimer cation radical $C_8H_{16}^{+\cdot}$ could lead to the propagation on the charge site. The doubly charged ions could result from the second ionization of the growing chains by removal of the odd electron. This takes place when the size of the growing chain becomes large enough to overcome the Coulomb repulsion. This leads to an interesting situation where the propagation takes place at both ends of the chain as shown below:

$$^+(\overset{|}{\underset{|}{C}} - CH_2)_n \overset{|}{\underset{|}{C}} - CH_2 - CH_2 - \overset{|}{\underset{|}{C}} (CH_2 - \overset{|}{\underset{|}{C}})_n^+$$

Note that the Coulomb repulsion between the two positive charges will be relaxed further as more isobutene monomer is added to the chain since the Coulomb potential varies inversely with the distance of charge separation. This situation is similar to gas phase polymerization of 1,3-butadiene initiated by the dications of C_{60} and C_{70} as observed by Bohme and coworkers.[31,32]

B. Metal - Catalyzed Systems. The majority of olefin polymerization processes are catalytic, most often involving a heterogeneous catalyst.[1,2,33-36] The study of the analogous processes in the gas phase and within clusters is important not only for probing the mechanism of the catalytic process and the exact nature of the catalyst-cocatalyst interaction, but also for the design of new catalysts with tailored reactivity and selectivity.

The reactions taking place following the process of metal ion generation in the presence of unsaturated hydrocarbons are of considerable interest because these reactions may provide a better understanding of related processes involving supported metal catalysts and the polymerization of olefinic monomers. For example, laser vaporization / ionization of metals can induce polymerization reactions as we recently reported for isobutene which is known to be polymerized in the bulk phase only by cationic mechanisms.[37-41] In these experiments, metal cations are produced in the gas phase by laser vaporization in the presence of isobutene vapor under a total pressure of 10^{-1} - 10^{-3} Torr. Other molecular and fragment ions from the isobutene monomer can also be generated by laser plasma ionization.[37-41]. These ions, including the metal cations, undergo several steps of bimolecular gas phase reactions before they eventually reach the liquid monomer phase where further reactions and higher polymerization steps take place. Unfortunately, in this experiment as well as in other bulk polymerization techniques, it is difficult to observe the propagation of the active species and to characterize the single steps of the polymerization reaction. Therefore, it is desirable to study the gas phase reactions induced by cations generated by the laser vaporization and plasma processes in order to better understand the mechanism of the liquid polymerization and identify the ions involved in the initiation and the early stages of propagation. We have demonstrated the study of catalyzed polymerization of isobutene clusters using the Al^+/t-BuCl system[14] and initiated other studies involving reactions with transition metal ions such as Ti^+, Zr^+, Zn^+, Pd^+ and Mo^+.[11,30] In the following sections, we present and discuss the results of the reactions between Ti^+, Zr^+ and Mo^+ with isobutene in the gas phase and within clusters.

Reactions of Ti^+ and Zr^+ with Isobutene Clusters.

The intracluster reactions of isobutene $(C_4H_8)_n$ following the interactions with Ti^+ were investigated and a typical clusters' mass spectrum is shown in Figure 6. The intact ion $Ti^+(C_4H_8)$ is not observed and the first product observed corresponds to a dehydrogenation reaction which generates the ion $Ti(C_4H_6)^+$. At lower laser power the relative intensity of the sequence $Ti(C_4H_6)(C_4H_8)_n^+$ increases at the expense of the $Ti(C_4H_8)_n^+$ series. Presumably the ions $Ti(C_4H_6)(C_4H_8)_n^+$ are generated through consecutive additions of C_4H_8 molecules onto $TiC_4H_6^+$ and it is expected to enhance these exothermic additions if the ion $TiC_4H_6^+$ does not possess much excess energy. This suggests that thermalized $TiC_4H_6^+$ ions should exhibit greater reactivity toward isobutene additions. Alternatively, the $Ti(C_4H_6)(C_4H_8)_n^+$ could represent the $TiC_4H_6^+$ cation surrounded by several $(C_4H_8)_n$ molecules. However, the significant decrease in the $TiC_4H_6^+$ ion intensity and the simultaneous increase in the ion intensities of larger $TiC_4H_6(C_4H_8)_n^+$ species under low laser power conditions are consistent with sequential reactions between $TiC_4H_6^+$ and C_4H_8 molecules. Furthermore, higher order elimination reactions producing species such as $TiC_{11}H_{18}^+$, $TiC_{10}H_{16}^+$ and $TiC_9H_{16}^+$ [corresponding to elimination of CH_4, C_2H_6 and C_3H_6; respectively from $Ti(C_4H_6)(C_4H_8)_2^+$] are observed. Since these ions are observed only under low laser power conditions, their generation via fragmentation of larger clusters is unlikely.

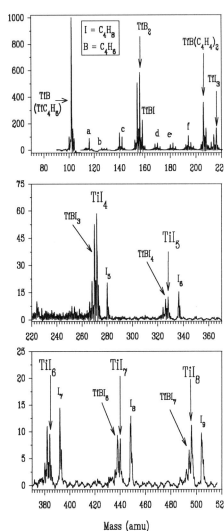

Figure 6. Mass spectrum for the Ti^+-Isobutene cluster system. Beam composition is 20% isobutene in He, $P_0=20$ psi, $T_0=298$ K. I=C_4H_8, B=C_4H_6, labeled peaks: a) TiBCH$_2$, b) [TiB$_2$-C$_2$H$_6$], c) TiBC$_3$H$_4$, d) TiB$_2$CH$_2$, e) [TiB$_3$-C$_2$H$_6$], f) [TiB$_3$-CH$_4$].

Another intriguing feature of Ti^+/isobutene clusters is the observation of the $(C_4H_8)_n^+$ and $H^+(C_4H_8)_n$ ions with $n \geq 5$ only under high laser power conditions (>20 mJ/cm^2). These ions can be formed as a result of dissociative charge transfer (DCT) within the unreactive series $Ti(C_4H_8)_n^+$ according to:

$$Ti^{+\cdot}.(C_4H_8)_n \xrightarrow{ CT } Ti.(C_4H_8)^+_n$$

$$\downarrow \text{Dissociation}$$

$$C_4H_9(C_4H_8)^+_{n-1} + {}^{\cdot}C_4H_7 \xleftarrow{ \text{Proton transfer} } (C_4H_8)^+_n + Ti$$

Similar to our interpretation of the $(C_4H_8)_n^{+\cdot}$ clusters, the species dissociated from the $Ti(C_4H_8)_n^+$ clusters are thought to have the structure (I):

$$- C(\!\!-\!CH_2\!-\!\overset{|}{\underset{|}{C}}\!\!\overset{+}{\!\!)_n} \cdots\!\!\cdots C_4H_7^{\cdot} \qquad (I)$$

which because of the slow evaporation of $C_4H_7^{\cdot}$ from the larger clusters, would appear as the intact series $(C_4H_8)_n^{+\cdot}$. This is consistent with the observation of $C_4H_9(C_4H_8)_n^+$ only for small n, where the evaporation of $C_4H_7^{\cdot}$ is more likely.

We also studied the gas phase reactions between Ti^+ and isobutene using LV-HPMS as described in the experimental section.[27] At low pressure, the dominant reaction path is the elimination of H_2 followed by sequential additions of isobutene molecules onto $TiC_4H_6^+$. At higher pressure (0.3 torr), the eliminative reactions are quenched and a new condensation channel is opened which corresponds to the generation of a polymeric ion of the form

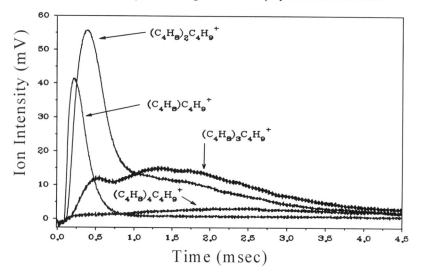

Figure 7. Temporal profiles of $C_4H_9^+(C_4H_8)_n$ at a source pressure of 0.3 torr isobutene.

$C_4H_9^+(C_4H_8)_n$. This channel proceeds via the generation of $C_4H_9^+$ from a pressure dependent chain transfer reaction of the Ti^+ containing ions to isobutene monomers.

The time profiles of the observed ions contain information on the relative rates of the sequential reactions. Figure 7 shows the temporal profiles of the $C_4H_9^+(C_4H_8)_n$ series measured at total source pressures of 0.3 torr. The time profile of the decay of $C_4H_9^+$ is essentially the same as that of $C_4H_9^+(C_4H_8)$ suggesting that the first two steps are very fast. However, the decay rates for the ions $C_4H_9^+(C_4H_8)_n$ with n = 3 - 5 are relatively slow. The strong dependence of the rate of cationic propagation on size can be explained in terms of steric or structural factors which may result in a situation where the charge becomes less accessible to the incoming monomers, thus slowing down further propagation steps.

In order to see if other transition metal ions would exhibit similar reactivity toward isobutene as Ti ions, we examined the reactions of Zr with isobutene clusters. Zr forms "cation-like" catalysts such as $CP_2ZrCH_3^+$ which are known catalysts for olefinic polymerization.[42,43] Figure 8 displays the mass spectrum observed following the interaction of Zr^+ with isobutene jet expansion. In addition to the $Zr(C_4H_6)(C_4H_8)_n^+$ series, the spectrum contains extensive dehydrogenation and elimination products. The general pattern observed corresponds to $ZrC_xH_y^-$ with x=4-16. In addition, several hydrocarbon ions such as C_4H_9, C_5H_9, $C_5H_9(C_4H_8)_n$ and $(C_4H_8)_n$ are observed. The generation of larger isobutene ions $(C_4H_8)_n^{+\cdot}$ is similar to the Ti^+ system and again these ions could have the structure (I).

Similar to the $C_4H_9^+$, the $C_5H_9^+$ can add isobutene molecules thus forming a growing ion of the type:

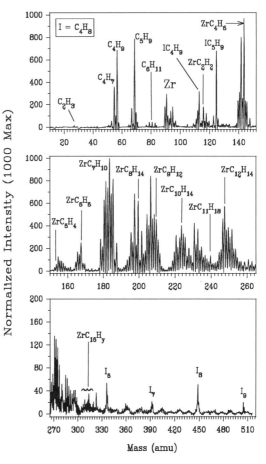

Figure 8. Mass spectrum for the Zr^+-Isobutene cluster system. Beam composition is 4% isobutene in He, P_0=22 psi, T_0=298 K.

$$H_2C=CH-\underset{|}{\overset{|}{C}}\left(CH_2-\underset{|}{\overset{|}{C}}\right)_{n-2}CH_2-\underset{|}{\overset{|}{C}}{}^+$$

In comparing the isobutene reactions with Ti^+ and Zr^+, it appears that Zr^+ is much more reactive thus leading to a very complicated mass spectrum. This reactivity can be attributed to the larger d-orbitals and the availability of f-orbitals which can contribute to both C-C and C-H bond insertion reactions.

Gas Phase reactions of Mo^+ with isobutene. The reactions between Mo^+ and isobutene have been studied using the LV-HPMS technique. Figures 9-a and 9-b display the mass spectra obtained following the generation of Mo^+ at two different pressures of isobutene. At low isobutene pressure, the main sequence of ions correspond to the addition of isobutene into the Mo^+ thus producing the $Mo^+(C_4H_8)_n$ series with n=1-4. Unlike the reactions with Ti^+, the dehydrogenation reactions producing $Mo^+(C_4H_{8-x})_n$ are very weak. However, the generation of $C_4H_9^+$ and $C_5H_9^+$ and their addition products on isobutene are also observed. It is interesting to note that the additions of the isobutene onto the t-butyl cation become quite dominant at higher isobutene pressures. Another interesting feature is the rapid shift in the ion distribution of the $Mo^+(isobutene)_n$ series toward higher n and the complete depletion of the ion intensities corresponding to n=1 and 2. Also, peaks corresponding to $Mo^+(C_4H_8)_3C_xH_y$ and $Mo^+(C_4H_8)_4C_xH_y$ are observed for x=1-3. This suggests the occurrence of sequential reactions between $Mo^+(C_4H_8)_3$ or $Mo^+(C_4H_8)_4$ and isobutene which result in elimination

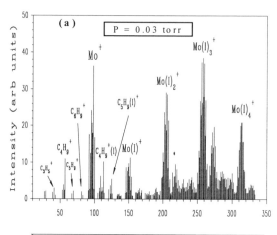

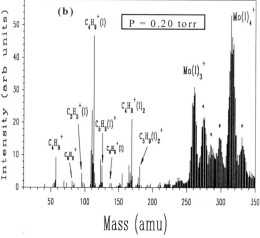

Fig 9. LV-HPMS mass spectrum of Mo^+/Isobutene at isobutene pressures of (a) 0.03 torr and (b) 0.2 torr, $I=C_4H_8$.

of radical species such as $CH_3\cdot$, $C_2H_5\cdot$ and $C_3H_7\cdot$. The time profiles for the major ions produced in the Mo^+/isobutene systems are shown in Figure 10 obtained from an experiment with 0.005 torr isobutene with a total source pressure of 0.495 torr using He as a carrier gas . The data confirms the sequential nature of the reactions observed and also suggests that the generation of the $C_4H_9^+$ takes place after the generation of the Mo^+ and almost parallel to the generation of $Mo(C_4H_8)_2^+$.

The sharp changes in the mass spectrum by increasing the pressure of isobutene can be explained by the sequential nature of the addition reactions. However, the complete disappearance of the Mo^+, $Mo^+(C_4H_8)$ and $Mo^+(C_4H_8)_2$ ions could be explained by a catalyzed reaction involving the $Mo^+(C_4H_8)_2$ ion according to:

$$Mo^+(C_4H_8)_2 + C_4H_8 \longrightarrow C_4H_9(C_4H_8)^+ + C_4H_7\cdot + Mo \qquad (4)$$

The occurrence of this reaction together with the shift of the sequential reactions toward higher adducts by increasing the pressure of isobutene can explain the observed depletion of the ion signals corresponding to Mo^+, $Mo^+(C_4H_8)$ and $Mo^+(C_4H_8)_2$.

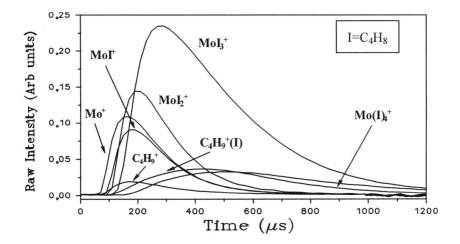

Figure 10. Temporal profiles for major ions produced following the reaction of Mo^+ ions with 0.005 torr isobutene with a total source pressure of 0.495 torr using He as a carrier gas.

IV. Summary and Implications

New fundamental insights on the details of polymerization reactions can be obtained by new methods for gas phase and cluster studies.[4-32,37-41] There are several reasons why these studies can provide detailed information that can't otherwise be obtained by conventional methods. In the gas phase and in clusters, one can observe the direct formation, in real time, of product polymeric ions of a chosen size. In the liquid phase, one is usually forced to infer what has happened by the qualitative and quantitative analysis of the products. Therefore, these solvent-free gas phase and cluster studies show promise for a fundamental

244

understanding of the elementary events occurring in the course of polymerization. In addition, they can lead to the discovery of new initiation mechanisms and for the synthesis of advanced materials such as defect-free, uniform polymeric films with excellent electrical and optical properties.

The study of the reactions of metal cations and olefins in the gas phase and within clusters can provide important information regarding the kinetics, energetics and structures of the adducts formed during metal-catalyzed polymerization reactions. Of particular interest are the reactions of the doubly charged cations such as Pd^{+2} and Ni^{+2} with the olefin monomers and clusters. These cations are often involved in catalyzed polymerization in the condensed phase.[33-36] Future work will address the reactions with doubly charged metal cations. This work will open the possibility to synthesize *in situ* gas phase catalysts for olefin polymerization by reacting the metal dications with appropriate ligands in the gas phase. This approach builds upon the recent work of Herschbach and his coworkers where a catalytic supersonic nozzle has been utilized to produce higher hydrocarbons from radical catalyzed reactions of C_2, C_3 and C_4 alkanes or alkenes.[44-46] It is expected that these experiments will provide a simple and well-controlled, solvent-free approach to the study of gas phase polymerization.

Acknowledgment

This work was supported by the National Science Foundation Grant CHE 9311643.

V. References

1. Kennedy, J. P.; Marechal, E. "Carbocationic Polymerization", John Wiley &Sons, New York (1982).
2. a- Odian, G. "Principles of Polymerization", McGraw-Hill, New York (1970).
   b- Faust, R.; Shaffer, T. D. (Eds) "Cationic Polymerization: Fundamentals and Applications" ACS Symposium Series 665, Washington DC (1997).
   c- Matyjaszewski, K (Ed.) "Cationic Polymerization: Mechanisms, Synthesis, and Applications" Marcel Dekker, Inc., New York (1996).
   d- Jenkins, A. D.; Ledwith, A. "Reactivity, Mechanism and Structure in Polymer Chemistry", John Wiley & Sons, London (1974).
3. Dalgarno, A. "The Chemistry of Astronomical Environments", J. Chem. Soc. Faraday Trans. **89**, 2111 (1993).
4. Castleman, A. W. Jr.; Wei, S. Annu. Rev. Phys. Chem. **45**, 685 (1994)
5. Castleman, A. W. Jr.; Bowen, K. H. Jr. J. Phys. Chem. **100**, 12911 (1996)
6. El-Shall, M. S.; Marks, C., J. Phys. Chem. **95**, 4932 (1991).
7. El-Shall, M.S.; Schriver, K. E., J. Chem. Phys. **95**, 3001 (1991).
8. Tsukuda, T.; Kondow, T., J. Chem. Phys. **95**, 6989-6992 (1991).
9. Coolbaugh, M. T.; Vaidyanathan, G.; Peifer, W. R.; Garvey, J. F., J. Phys. Chem, **95**, 8337 (1991).
10. El-Shall, M. S. in "The Physics and Chemistry of Finite Systems: From Clusters to Crystals", Eds; Jena, P., Rao, B. K., Khanna, S. Vol.II, 1083, Kluwer Academic Publishers, 1992.
11. Daly, G. M.; El-Shall, M. S., Z. Phys. D. **26S**, 186 (1993).
12. Tsukuda, T.; Kondow, T., J. Am. Chem. Soc. **116**, 9555-9564 (1994).
13. Desai, S. R., Feigerle, C. S., Miller, J. J. Phys. Chem. **99**, 1786 (1995).
14. Daly, G. M.; El-Shall, M. S., J. Phys. Chem., **99**, 5283 (1995).
15. Daly, G. M.; Y. B. Pithawalla, Y. B.; Yu, Z.; El-Shall, M. S., Chem. Phys. Letters , **237**, 97-105 (1995).
16. El-Shall, M. S.; Daly, G. M.; Yu, Z.; Meot-Ner, M., J. Am. Chem. Soc., **117**, 7744-7752 (1995).

17. El-Shall, M. S.; Yu, Z., J. Am. Chem. Soc., **118**, 13058-13068 (1996).
18. Pithawalla, Y. B.; J. Gao, J.; Yu, Z.; El-Shall, M. S. Macromolecules, **29**, 8558-8561 (1996).
19. Tsukuda, T.; Kondow, T.; Dessent, C. E. H.; Bailey, C. C.; Johnson, M. A.; Hendricks, J. H.; Lyapustina, S. A.; K. H. Bowen, K. H. Chem. Phys. Letters, **269**, 17-21 (1997).
20. Zhong, Q., Poth, L., Shi, Z., Ford, J. V., Castleman, Jr., A. W. J. Phys. Chem. **101**, 4203-4208 (1997).
21. Raksit, A. B.; Bohme, D. K. Can. J. Chem. **62**, 2123 (1984).
22. Forte, L.; M. H. Lien, M. H.; Hopkinson, A. C.; Bohme, D. K., Makromol. Chem. Rapid Commun. **8**, 87 (1987).
23. Forte, L.; Lien, M. H.; Hopkinson, A. C.; Bohme, D. K., Can. J. Chem. **67**, 1576 (1989).
24. Guo, B. C.; Castleman, A. W., Jr., J. Am. Chem. Soc. **114**, 6152 (1992).
25. Brodbelt, J. S.; Liou, C. C.; Maleknia, S.; Lin, T. J.; Lagow, R. J., J. Am. Chem. Soc **115**, 11069 (1993).
26. Bjarnason, A.; Ridge, D. P. J. Phys. Chem., **100**, 15118-15123 (1996).
27. Daly, G. M.; El-Shall, M. S., J. Phys. Chem., **98**, 696 (1994).
28. Meot-Ner, M.; Sieck, L. W.; El-Shall, M. S.; Daly, G. M. J. Am. Chem. Soc., **117**, 7737-7743 (1995).
29. Meot-Ner, M.; Pithawalla, Y. B.; Gao, J.; El-Shall, M. S. J. Am. Chem. Soc., **119**, 8332-8341 (1997).
30. El-Shall, M. S. "From Gas Phase Clusters to Polymers: A Route to Novel Materials in "Science and Technology of Atomically Engineered Materials", edited by Jena, P; Rao, B. K.; Khanna, S. World Scientific (1996), pp. 67-75.
31. Wang, J.; Javahery, G.; Petrie, S.; Bohme, D. K., J. Am. Chem. Soc. **114**, 9665 (1992).
32. Baranov, V.; Wang, J.; Javahery, G.; Hopkinson, A. C.; Bohme, D. K. J. Am. Chem. Soc., **119**, 2040 (1997).
33. Jiang, Z.; Sen, A. J. Am. Chem. Soc. **112**, 9655-9657 (1990).
34. Reddy, N. P.; Yamashita, H.; Tanaka, M. J. Am. Chem. Soc. **114**, 6596-6597 (1992).
35. Johnson, L. K.; Killian, C. M.; Brookhart, M. J. Am. Chem. Soc. **117**, 6414-6415 (1995);
36. Johnson, L. K.; Mecking, S. ; Brookhart, M. J. Am. Chem. Soc. **118**, 267-268 (1996).
37. Vann, W.; El-Shall, M. S., J.Am. Chem. Soc. **115**, 4385 (1993).
38. Vann, W.; Daly, G. M.; El-Shall, M. S. in "Laser Ablation in Materials Processing: Fundamentals and Applications", Braren, B.; Dubowski, J. and Norton, D. Eds. Materials Research Society Symposium Proceedings Series, **285**, 593 (1993).
39. Deakin, L.; Den Auwer, C.; Revol, J. F.; Andrews, M. P. J. Am. Chem. Soc. **117**, 9916 (1995).
40. El-Shall, M. S.; Slack, W., Macromolecules, **28**, 8456 (1995).
41. El-Shall, M. S. Applied Surface Science, **106**, 347 (1996).
42. X. Yang, C. L. Stern and T. J. Marks, J. Am. Chem. Soc. **1991**, 113, 3623; H. Kawamura-Kuribayashi, N. Koga and K. Morokuma, J. Am. Chem. Soc. **1992**, 114, 2359 ; C. A. Jolly and D. S. Marynick, J. Am. Chem. Soc. **1989**, 111, 7968.
43. C. S. Christ, J. R. Eyler and D. E. Richardson, J. Am. Chem Soc **1988**, 110, 4038; **1990**, 112, 597.
44. Shebaro, L.; Bhalotra, S. R.; Herschbach, D. J. Phys. Chem. A **1997,** 101, 6775.
45. Shebaro, L.; Abbott, B.; Hong, T.; Slenczka, A.; Friedrich, B.; Herschbach, D. Chem. Phys. Lett. **1997,** 271, 73.
46. Kim, S. K.; Lee, W.; Herschbach, D. J. Phys. Chem. **1996,** 100, 7933.

MISCELLANEOUS SOLVENT-FREE POLYMERIZATIONS AND PROCESSSES

Chapter 16

Solvent-Free Functionalization of Polypropylene by Reactive Extrusion with Acidic Peroxides

S. C. Manning and R. B. Moore

Department of Polymer Science, University of Southern Mississippi, P.O. Box 10076, Hattiesburg, MS 39406–0076

Variable quantities of functionalized peroxides bearing carboxylic acid groups were reacted with polypropylene (PP) in a twin screw extruder. Systematic variations in the molecular structure of the peroxides were found to significantly affect the efficiency of grafting the carboxylic acid groups onto PP and the polymer degradation process. This behavior was attributed to the relative reactivities of the different free radicals generated by thermal decomposition of the peroxides. Furthermore, the functionalized polypropylene (f-PP) was investigated as a compatibilizing additive for 80/20 PP/PA-6,6 (polyamide 6,6) blends. With incorporation of the f-PP into the blends, differential scanning calorimetry (DSC) showed an 80°C decrease in the PA-6,6 crystallization temperature. A near linear increase in the impact strength of the blends was observed with f-PP incorporations up to 30% of the PP phase. Blends containing 30% f-PP demonstrated impact properties approaching that of pure PA-6,6.

As a result of recent concerns over the environmental hazards of volatile organic compounds (VOCs), many solvent-free polymerizations and processes have been developed. For example, reactive extrusion($1,2$) has emerged as an ideal low VOC technique that involves melt-phase chemical reactions performed within the configuration of an extruder. The ability to alter the down stream configuration of the extruder allows for the combination of several chemical procedures into one continuous process. Furthermore, as compared to typical solvent born batch reactors, most reactive extrusion procedures are based on a twin screw geometry(3) which creates a melt mixed environment with excellent heat and mass transfer properties.

Typical reactive extrusion processes may be controlled through tailored zone architecture (i.e., variable screw design, temperature profiles, etc.) and through the rate and position at which the reactants are introduced into the extruder. Figure 1 is a schematic diagram (side view) of a twin-screw extruder illustrating variability in

hopper

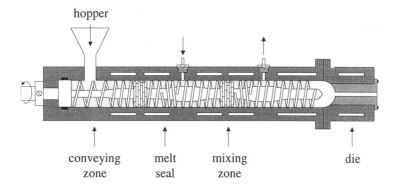

conveying   melt   mixing   die
zone   seal   zone

Figure 1. Side-view schematic of a twin-screw extruder.

the screw design. Intense mixing zones may be incorporated into the screw design through the addition of kneading blocks. Furthermore, melt seals may be created through a reverse flight geometry; these seals are often used to isolate individual reaction zones along the extruder path. Strategic placement of through-barrel ports allows for injection of reactants for sequential reaction initiation as well as devolatilization of residual monomers, condensates, and/or other by-products.

Within the field of reactive extrusion, organic peroxides offer the advantage of a thermally induced decomposition by homolytic cleavage of the labile O-O bonds. This cleavage forms two free radical species that may be able to abstract hydrogen atoms from polymer chains and/or add to double bonds. Furthermore, organic peroxides are generally required in low concentrations and can decompose to complete the chemical reaction during typical extrusion residence times. As a result of these desirable characteristics, a wide variety of organic peroxides have been commercialized for reactive extrusion applications such as: bulk polymerization,(4,5) cross-linking of polyethylene and elastomers,(6) production of controlled rheology polypropylene,(7-9) and grafting of functional monomers onto polymer chains.(10-14)

Organic peroxides are ideally suited for applications involving modifications of polymer properties via melt-phase chemical reactions. To optimize the chemical compatibility and/or solubility of the peroxides in a wide variety of polymeric systems, the organic character of these peroxides may be tailored by subtle changes in the molecular structure. Moreover, since applied extrusion conditions (e.g., melt temperatures and zone residence times) vary with polymer type, the numerous structures available with organic peroxides yield a wide range of decomposition kinetics and resulting free radical reactivities.

Functionalization of polyolefins is often achieved by grafting a polar monomer such as maleic anhydride (MAH) onto the polyolefin backbone using reactive peroxides.(12-17) Compared to other initiation techniques (e.g., γ-irradiation), the use of organic peroxides initiators for MAH grafting onto PP generally results in superior MAH grafting efficiencies.(18) This functionalization reaction can be performed in solution(19) or in the melt,(17) and may be an economic way to make these inherently nonpolar polymers more compatible with polar polymers. For example, enhanced minor phase dispersion and interfacial adhesion has been observed with blends of MAH-functionalized polypropylene with nylon 6(20) and when MAH-functionalized polypropylene was used as a compatibilizing agent in blends of polypropylene with nylon 6,6.(21)

During a typical melt graft reaction of MAH onto polypropylene, the reactive radicals (i.e., the decomposition products from the peroxide initiators) abstract hydrogens from the polypropylene backbone to form tertiary radicals along the chains. These polymeric radicals then add to the reactive double bond of MAH to form randomly distributed grafts.(12) While the initial product of this free radical addition reaction is another radical, recent ^{13}C NMR studies of MAH grafted PP(22) have shown that the graft sites of the final product contain predominately single succinic anhydride rings. The possibility for continued MAH homopolymerization into block side chains has been suggested to be unlikely based on ceiling temperature considerations(23) (i.e., at typical melt temperatures of ca. 180° C); however, a portion of the grafted MAH groups may form cross-links between polypropylene chains.(24)

In addition to the desirable grafting reactions, polypropylene has also been shown to degrade during peroxide-initiated functionalization by tertiary radical β-scission (i.e., chain scission at the site of the polymeric radicals). At low MAH concentrations, the secondary anhydride radicals have been found to contribute to the formation of tertiary polymeric radicals via intramolecular chain transfer;(14) this rearrangement consequently promotes polypropylene degradation.(12,14)

Recently, Elf Atochem N.A. has developed a series of asymmetric, functional peroxides (peroxyesters) bearing carboxylic groups which may be used to graft acidic functionality directly onto polyolefin chains.(25) In addition, DeNicola has demonstrated that unsaturated peroxyacids may be used to functionalize polypropylene in order to increase the interfacial adhesion to glass fiber reinforcement.(26) The original hypothesis in the use of these new peroxides in reactive processing is that highly reactive alkyl radicals from the thermal decomposition of the asymmetric peroxides abstract hydrogens from polypropylene. Subsequently, less reactive radicals (containing the acid functionality) couple with the polymeric radicals to form the grafts. Since these grafts are created during a radical termination step rather than in a chain mechanism, undesirable cross-linking and/or β-scission may be minimized.

In this chapter, we compare the grafting efficiency of a series of carboxylic acid containing peroxides in a reactive extrusion process with polypropylene (PP). The focus of this study is on the subtle changes in the chemical structures of these peroxides that differ by only one methylene unit and/or one site of unsaturation. The grafting mechanism is also correlated to other possible free radical mechanisms such as β-scission and cross-linking. Furthermore, the resulting f-PP will be investigated as a compatibilizing agent in PP/PA-6,6 blends.

Experimental Section

Materials. Polypropylene powder (Pro-Fax 6501) was obtained from Himont Inc. Polyamide 6,6 (Vydine 21) was obtained from Monsanto. The functional peroxides Luperox PMA, Luperox TA-PMA, Luperco 212-P75 and Lupersol 512 were obtained from Elf Atochem North America. Luperco 212-P75 was obtained and used as a powder of the peroxide on a polypropylene carrier; Lupersol 512 was obtained as a 50 wt% solution in ethyl 3-ethoxypropionate and evaporated onto a PP powder carrier. The structures of these peroxides are represented in Figure 2.

Reactive Extrusion. Polypropylene and the peroxide/PP powders were premixed to obtain samples of 0.5, 1.0 and 2.0 wt% peroxide. The mixture was then extruded in a Haake Rheochord 90 counter-rotating twin screw extruder with 3 mixing zones and a rheological slit die. The temperatures of the zones were $T_1=150^\circ C$, $T_2=T_3=180^\circ C$, T_4 (die)=$170^\circ C$. The screw speed was set at 15 RPM corresponding to a residence time of ca. 7 minutes, and all samples were extruded under a N_2 atmosphere.

Blends of PP and PA-6,6 were premixed to obtain an overall 80/20 (w/w) composition of PP/PA-6,6. Polypropylene functionalized with 0.5% Luperox PMA was incorporated into the PP portion of the blends at 5, 10, 20 and 30 wt%. The premixed blends were extruded with a temperature profile of $T_1 = T_2 = T_3 = T_4 = 270^\circ C$. The screw speed was set at 15 rpm, and all blends were extruded under a N_2 atmosphere. Physical testing samples were obtained by injection molding the blend regrind at $270^\circ C$ using a Boy 15S injection molder.

Titrimetric Assay of the Functionalized Polypropylene. To eliminate the potential contribution of unreacted peroxide, the extruded products were first purified by reprecipitation. The extrudate was dissolved in xylene at $130^\circ C$ to a concentration of 5% (w/v) and reprecipitated into methanol. After filtration, the precipitate was washed with pure methanol and dried at 60 $^\circ C$ in vacuo. Samples of the purified polypropylene (0.5 g) were dissolved in 100 ml of xylene and titrated to the phenolphtalein endpoint with a standardized solution of benzyltrimethylammonium hydroxide in MeOH. The titrant was standardized with benzoic acid in xylene. All titrations were performed in triplicate at $110^\circ C$ under a N_2 atmosphere. Due to potential oxidation of PP during melt extrusion, the residual acidity of pure PP (extruded under identical conditions) was assayed as a blank and subtracted from the titrimetric results of the functionalized PP samples. The acid contents and sample standard deviations were calculated in units of equivalents per gram of polymer. The grafting efficiency for the peroxides were then calculated as:

$$\text{Grafting Efficiency \%} = \frac{equiv.\ f\text{-}PP\ -\ equiv.\ PP}{equiv.\ Peroxide\ in\ Feed} \times 100 \qquad (1)$$

where *equiv. f-PP* is the total acidity of the functionalized PP, *equiv. PP* is the blank acidity, and *equiv. Peroxide in Feed* is the initial concentration of the peroxide/PP mixture.

Characterization of Material Properties. The purified polypropylene samples were densified by melting at $190^\circ C$ in vacuo and then reground into a coarse powder. The Melt Index of the products was measured with a Custom Scientific Instruments CSI Melt Indexer at $230^\circ C$ and under a 2.16 kg load as per ASTM D-1238. The molecular weights of the PP samples were measured with high-temperature size exclusion chromatography (SEC) using a differential refractive index detector. The samples

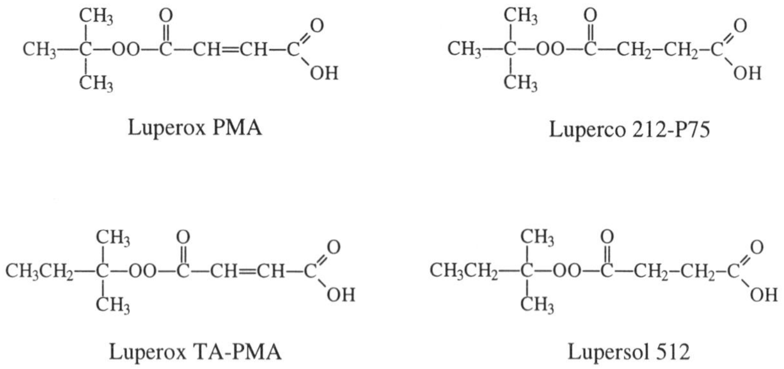

Luperox PMA

Luperco 212-P75

Luperox TA-PMA

Lupersol 512

Figure 2. Chemical structures of asymmetric functional peroxides.
(Reproduced with permission from reference 25. Copyright 1997 Society of
Plastics Engineers.)

were dissolved in filtered 1,2,4-trichlorobenzene (TCB) at 145°C. A 0.1 weight
percent antioxidant (butylate hydroxy-toluene) was used to stabilize the samples. The
SEC column set was heated to 145°C and calibrated using 25 polystyrene standard
samples with molecular weights ranging from 950 to 15 x 10^6 g/mol. The polystyrene
calibration curve was converted to one for PP using Mark-Houwink coefficients for
polystyrene and PP in TCB.(26)

An Electroscan scanning electron microscope was used to verify the phase
dimensions of the PP/PA-6,6 blends. Micrographs samples were prepared by freeze
fracturing runners from the injection molded components in liquid N_2. The thermal
transitions of the blends were monitored with a Perkin Elmer DSC 7. Both heating
and cooling thermograms were obtained between 50°C and 300°C at a scanning rate
of 20°C/min, under a N_2 atmosphere.

Notched Izod impact tests were performed with a Custom Scientific
Instrument CSI 137-053 as per ASTM D-256. Impact strengths were taken as an
average of 15 samples. Tensile tests were performed on molded dog bone samples
with a Material Test System MTS 810 coupled with a microconsole MTS 458-20.
All tensile properties were taken as an average of 15 dog bones.

Results and Discussion

The series of the functional peroxides used in this study (see Figure 2) was chosen in
order to investigate the effect of systematic changes in the chemical structures on the
ability to graft carboxylic acid groups onto a polyolefin backbone. While each of the
asymmetric peroxides contain a single carboxylic acid group, they differ in aliphatic
character. A cross comparison of the four peroxides highlights two distinct structural
differences. Luperox PMA and Luperco 212-P75 differ from Luperox TA-PMA and
Lupersol 512 by one methylene unit. This differentiation will be used to compare the
effect of a t-butyl group versus a t-amyl group on the resulting free radical reactivity.
On the other hand, the pair of Luperox peroxides differ from Luperco 212-P75 and

Lupersol 512 by a single site of unsaturation. This differentiation will be used to compare the effect of a reactive double bond on the ultimate grafting process. As will be discussed below, these subtle chemical differences affect the respective radical reactivities and yield significant contrast in grafting efficiency during reactive extrusion with polypropylene.

Thermal Decomposition of the Functional Peroxides. At 180°C (i.e., the temperature of the melt-mixing zones in the extruder), the thermal decomposition half-times of the two Luperox peroxides are approximately the same ($t_{1/2}$ = ca. 0.27 min.) and only slightly less than that of the Luperco and Lupersol peroxides ($t_{1/2}$ = ca. 0.31 min.). Nevertheless, based on these $t_{1/2}$ data and given the 7 min. residence time in the extruder, all of the peroxides are assumed to decompose completely by the time the melt enters the rheological slit die. Thus, any differences in the overall grafting efficiencies are not considered to be a simple function of the peroxide decomposition kinetics.

The thermal decomposition of the peroxides proceeds by the general reaction shown in Scheme 1. As a result of the asymmetric structures of the functionalized peroxides, two free radicals with distinctly different reactivities are formed by the homolytic cleavage of the peroxide linkage. Note that this reaction includes a number of possible rearrangement products which are a consequence of free radical stabilization processes. Throughout the context of this paper, the initiating radicals (**A** through **D**) in Scheme 1 will be referred to as $R_1\bullet$ radicals and the coupling radicals bearing the acid functionality (**F** through **I**), will be referred to as $R_2\bullet$ radicals.

In order to understand the role of these decomposition products in the ultimate grafting process, it is necessary to first rationalize the relative reactivities of each of the radical species. For the t-alkyloxy radicals, **B** and **D**, rearrangement by β-scission and elimination of a molecule of acetone is also possible. However, the kinetics of this β-scission reaction differ significantly between the t-butoxy, **B**, and t-amyloxy, **D**, radicals.(27) The t-butoxy radicals can slowly rearrange to yield methyl radicals, **A**, of about the same reactivity as the t-butoxy radicals (corresponding bond dissociation energies of 105 and 104 kcal/mol, respectively). Furthermore, both the t-butoxy and methyl radicals are quite capable of abstracting hydrogen atoms off of the PP backbone. In contrast to the t-butoxy radicals, there is a strong thermodynamic driving force for β-scission of the t-amyloxy radicals.(27) This stabilization process yields an ethyl radical, **C**, which is a weaker hydrogen abstractor relative to the methyl radical as suggested by the lower corresponding bond dissociation energy of 98 kcal/mol.

For the decomposition products containing the R_2 moiety, spontaneous decarboxylation will occur if the carboxy radical is thermodynamically unstable relative to the functionalized alkyl, **G**, or alkenyl, **I**, radicals. Based on bond dissociation energy considerations, decarboxylation is assumed to occur for all of the saturated carboxy radicals, **F**, used in this study. However, due to the relatively high energy of the alkenyl radical (corresponding bond dissociation energy of ca. 104 kcal/mol), the thermodynamic driving force for decarboxylation of the 2-carboxy acryloyloxy radical, **H**, is significantly lower compared to that of the **F** radical. Therefore, alkenyl radicals and 2-carboxy acryloyloxy radicals are both expected to participate in the reactive process. The functionalized alkyl radicals are expected to be low energy radicals (ca. 98 kcal/mol) which mainly participate in coupling reactions

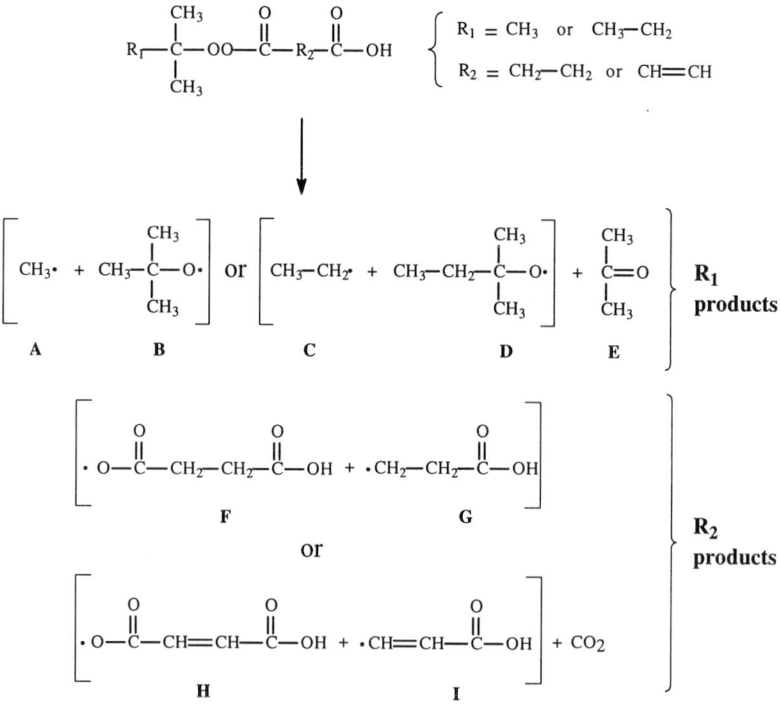

Scheme 1. Asymmetric functional peroxide decomposition scheme.

with the generated polymeric radicals, while the alkenyl and 2-carboxy acryloyloxy radicals possess reactivities which are capable of hydrogen abstraction as well as coupling reactions.

Formation of the Functional Graft. In light of the relative reactivities of the radicals in Scheme 1, it is reasonable to expect that the overall efficiency of grafting carboxylic acid groups onto PP will be governed by: 1) the ability of radical species to abstract hydrogens from the PP backbone and 2) the ability of the functionalized radicals to couple to the new polymeric radicals. One possible grafting mechanism is shown in Scheme 2 as a two step process of hydrogen abstraction by the $R_1\bullet$ radical (i.e., an initiation step) followed by a single radical coupling between $R_2\bullet$ and the polymeric radical (i.e., the termination step) to form the functionalized graft.

Once the polymeric radicals are formed by hydrogen abstraction, it is important to note that several other radical reactions are possible which do not lead directly to graft formation. For example, in the presence of free radicals, PP is known to degrade via β-scission(7-9,28) to yield terminal double bonds and terminal free radicals. Under the conditions of reactive extrusion, the β-scission of PP is expected to be a parallel reaction with the coupling reaction shown in Scheme 2.

While all of the peroxides used in this study yield radicals capable of coupling to the polymeric radical (as per Scheme 2), it is important to note that the <u>unsaturated</u> $R_2\bullet$ radicals from the two Luperox peroxides are further capable of both H-abstraction and free radical addition reactions. Scheme 3 illustrates a possible mechanism for grafting the functionalized alkenyl radical onto PP in the absence of the $R_1\bullet$ initiation step shown in Scheme 2. First, the functionalized alkenyl radical abstracts a hydrogen from PP to yield the polymeric radical. Then, within the same "melt-cage," the polymeric radical adds across the vinyl group of the newly formed acrylic acid molecule to form the desired acid graft. Note that while this mechanism is shown specifically with radical **I** it is likely that radical **H** could participate in a similar manner; this would yield a diacid graft.

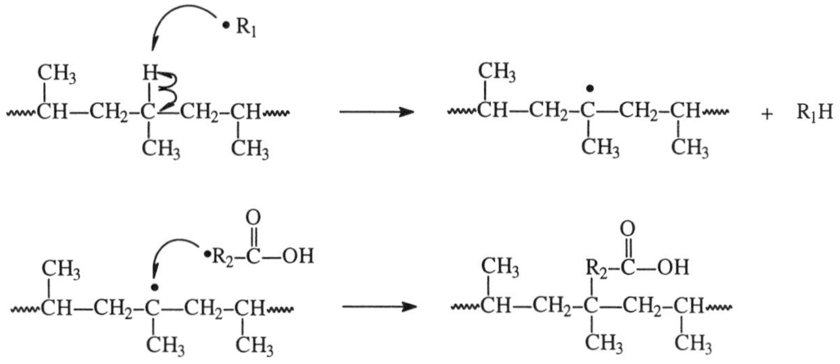

$$\text{CH}_3 \quad \text{H} \bullet R_1$$
$$\text{wwCH–CH}_2\text{–C–CH}_2\text{–CHww} \quad \longrightarrow \quad \text{wwCH–CH}_2\text{–C}^\bullet\text{–CH}_2\text{–CHww} \quad + \quad R_1\text{H}$$
$$\text{CH}_3 \quad \text{CH}_3 \qquad\qquad \text{CH}_3 \quad \text{CH}_3$$

$$\bullet R_2\text{–}\overset{\text{O}}{\overset{\|}{\text{C}}}\text{–OH}$$
$$\text{CH}_3 \quad {}^\bullet \qquad\qquad \text{CH}_3 \quad R_2\text{–}\overset{\text{O}}{\overset{\|}{\text{C}}}\text{–OH}$$
$$\text{wwCH–CH}_2\text{–C–CH}_2\text{–CHww} \quad \longrightarrow \quad \text{wwCH–CH}_2\text{–C–CH}_2\text{–CHww}$$
$$\text{CH}_3 \quad \text{CH}_3 \qquad\qquad \text{CH}_3 \quad \text{CH}_3$$

Scheme 2. H-abstraction followed by coupling reaction.
(Reproduced with permission from reference 25. Copyright 1997 Society of Plastics Engineers.)

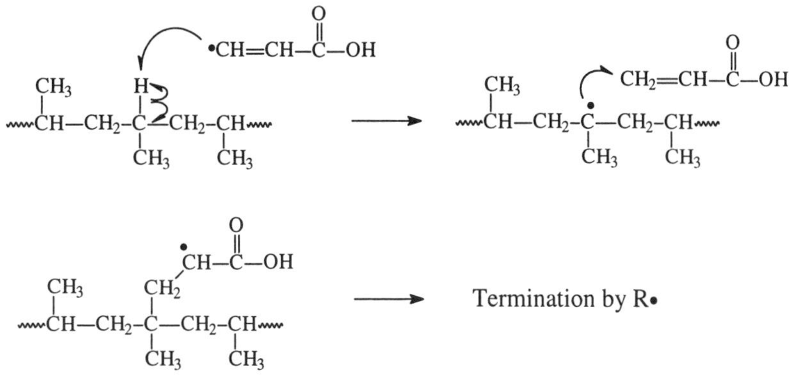

$$\bullet\text{CH}=\text{CH–}\overset{\text{O}}{\overset{\|}{\text{C}}}\text{–OH}$$
$$\text{CH}_3 \quad \text{H} \qquad\qquad \text{CH}_3 \quad \text{CH}_2=\text{CH–}\overset{\text{O}}{\overset{\|}{\text{C}}}\text{–OH}$$
$$\text{wwCH–CH}_2\text{–C–CH}_2\text{–CHww} \quad \longrightarrow \quad \text{wwCH–CH}_2\text{–C–CH}_2\text{–CHww}$$
$$\text{CH}_3 \quad \text{CH}_3 \qquad\qquad \text{CH}_3 \quad \text{CH}_3$$

$$\overset{\bullet}{\text{CH–}}\overset{\text{O}}{\overset{\|}{\text{C}}}\text{–OH}$$
$$\text{CH}_3 \quad \text{CH}_2 \qquad\qquad$$
$$\text{wwCH–CH}_2\text{–C–CH}_2\text{–CHww} \quad \longrightarrow \quad \text{Termination by R}\bullet$$
$$\text{CH}_3 \quad \text{CH}_3 \qquad\qquad$$

Scheme 3. H-abstraction followed by addition reaction.

Effect of Peroxide Structure on Reactive Functionalization. Figure 3 shows the titrimetric results of the functionalized polypropylene (*f*-PP) after reactive extrusion and sample purification. For each peroxide used, there is a slight increase in the acid content of the *f*-PP samples with an increase in the peroxide concentration (i.e., feed concentration, wt %). Moreover, the four peroxides yield significant differences in the quantity of acidic groups grafted onto the polypropylene. These differences in acidities can be attributed to the distinct contrast in relative reactivities of the radicals originating from the different chemical structures of the peroxides.

Of the four peroxides used in this study, the highest acidity was obtained with Luperox PMA. As discussed above, this peroxide yields both $R_1\bullet$ and $R_2\bullet$ radicals with the highest relative reactivities. The $R_1\bullet$ methyl radicals of Luperox PMA are very efficient at producing polymeric radicals, while the $R_2\bullet$ radicals are capable of grafting onto PP by Scheme 2 and/or Scheme 3. The lowest acidity of *f*-PP (Figure 3) was obtained with Lupersol 512. In contrast to Luperox PMA, this peroxide yields both $R_1\bullet$ and $R_2\bullet$ radicals with the lowest relative reactivities. The $R_1\bullet$ ethyl radicals of Lupersol 512 are inefficient at producing polymeric radicals, and the $R_2\bullet$ species are likely to graft onto PP only by the process shown in Scheme 2.

The intermediate acidities observed with Luperox TA-PMA and Luperco 212 can be attributed to the peroxides yielding a combination of $R_1\bullet$ and $R_2\bullet$ radicals with high and low relative reactivities. For example, Luperco 212 yields reactive methyl $R_1\bullet$ radicals and less reactive $R_2\bullet$ radicals. This peroxide is the ideal case for grafting by Scheme 2, and actually demonstrates the principle of the original hypothesis of this work. In comparing the observed acidity from Luperco 212 to that from Luperox TA-

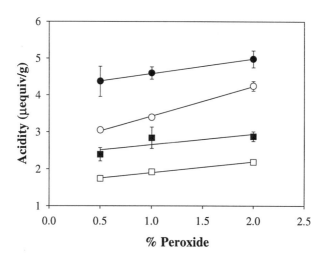

Figure 3. Titrimetric results indicating the acid content of the *f*-PP verses the peroxide concentrations in the feed: (●) Luperox PMA; (○) Luperox TA-PMA; (■) Luperco 212-P75; (□) Lupersol 512.
(Reproduced with permission from reference 25. Copyright 1997 Society of Plastics Engineers.)

PMA, it is important to note that the $R_1\bullet$ radical of Luperox TA-PMA is less reactive than the methyl radical from Luperco 212; however, Luperox TA-PMA yields the very reactive alkenyl $R_2\bullet$ radical. Since Luperox TA-PMA yields an acidity which is significantly greater than that observed with Luperco 212, these data suggest that the grafting process shown in Scheme 3 is not only likely, but that the reactivity of the $R_2\bullet$ radical is quite important in the overall grafting process.

Figure 4 compares the overall grafting efficiency of the four peroxides with respect to the feed concentration. For each feed concentration, the relative grafting efficiencies follow the trends in acidity (Figure 3); the greater the relative radical reactivities, the higher the grafting efficiency. The systematic decrease in grafting efficiency is correlated to the low slopes observed in the plots of Figure 3. A significant increase in peroxide feed concentration only yields a slight increase in acidity, and thus the grafting efficiency (as calculated with equation 1) diminishes at high peroxide concentrations.

The data in Figure 4 suggest that an increase in the feed concentration leads to high levels of ungrafted R_2 species. This behavior may be attributed to a combination of limited peroxide solubility (in the polypropylene melt)(29) and/or to a loss of low molecular weight f-PP during the sample purification procedure (see Experimental section). From group contributions(30), the solubility parameters of the peroxides were calculated to be 16.7, 16.7, 14.1, and 14.3 $(MPa)^{1/2}$ for Luperox PMA, Luperox TA-PMA, Luperco 212, and Lupersol 512, respectively. Due to a lack of correlation, these slight differences among the solubility parameters can not account for the trends in grafting efficiency at a given concentration. However, with respect to the solubility

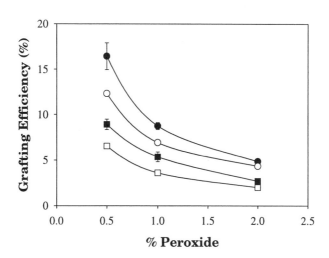

Figure 4. Grafting Efficiency for the peroxides as a function of peroxide concentration in the feed: (●) Luperox PMA; (○) Luperox TA-PMA; (■) Luperco 212-P75; (□) Lupersol 512.

parameter of PP ($\delta \approx 19.0$ (MPa)$^{1/2}$), it is reasonable to expect rather poor peroxide solubilization; this behavior would thus limit the amount of peroxide available for grafting.

The possible link between sample purification and the drop in grafting efficiency with increased peroxide concentration (see Figure 4) stems from potential β-scission degradation reactions of PP in the presence of free radicals.(*7-9,28*) With this chain scission reaction, it is possible that a significant fraction of low MW PP would be produced. Moreover, if these low MW chains contain grafted carboxylic acid sites, then it is likely that these short *f*-PP chains would not readily precipitate in methanol. Therefore, the titrimetric results may be more representative of grafting onto high MW *f*-PP chains. To test for this possibility, a sample of *f*-PP (prepared with 0.5 wt% Luperox PMA) was reprecipitated following the purification procedure described in the Experimental section. The filtrate was evaporated and found to contain ca. 2% of the original sample mass. FTIR analysis of this filtrate sample (after soxhlet extraction with water) indicated the presence of polypropylene with a high acid content. Titrations were performed on the recovered highly functionalized PP and revealed an average acidity of 350 meq/g which corresponded to an overall grafting efficiency for Luperox PMA of ca. 40%. Therefore, as the peroxide concentration in the feed increases, more chain scission reactions occur, more low MW *f*-PP is lost in the purification step, and thus the apparent grafting efficiency decreases.

A change in the molecular weight distribution of the extruded samples is evident in the GPC data of Figure 5. Relative to the pure PP extruded under identical conditions, each of the peroxides caused a significant decrease in the overall molecular weight. Specifically, it is important to note that the z-average molecular weight, M_z, which reflects the high molecular weight tail of the distribution (Figure 5A), is most sensitive to the radical induced chain-scission reactions. These results are consistent with previous studies of controlled molecular weight reduction of PP by the use of peroxides.(*7,9*) Furthermore, in agreement with the model of Suwanda et al.,(*9*) the more reactive methyl radicals from Luperco 212, as compared to the ethyl radicals from Lupersol 512, yielded the highest degree of molecular weight reduction by chain-scission. In contrast, the high molecular weight tail (as indicated by the M_z data) of the *f*-PP sample prepared with Luperox PMA was the least affected. Although Luperox PMA yields the reactive methyl radicals, the efficient grafting of the unsaturated $R_2\bullet$ radicals (as discussed above) presents the possibility of further cross-linking reactions between the unsaturated grafts and other polymeric radicals. These cross-linking reactions would thus be expected to maintain a significant presence of high molecular weight polymer.

The effect of changes in the molecular weight distribution of the *f*-PP samples during the reactive extrusion process is further evident by comparing the melt indices of the extruded products. Figure 6 shows clear differences in the melt indices of the *f*-PP samples prepared with the four peroxides. In agreement with the GPC data in Figure 5, the *f*-PP sample prepared with Luperco 212 yields the highest melt index. As the $R_1\bullet$ radical is changed to the less reactive ethyl species (i.e., with Lupersol 512), less chain-scission is induced, and the melt index drops.

The two Luperox peroxides yield the lowest melt indices (Figure 6). Again, due to less chain-scission, the peroxide which yields the weakest $R_1\bullet$ radical (i.e.,

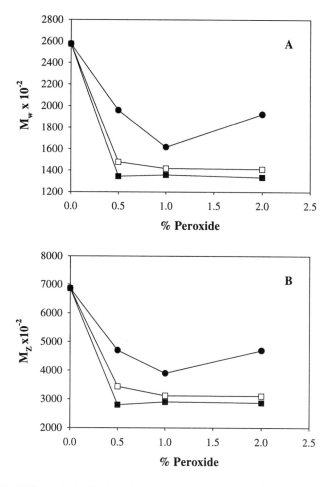

Figure 5. SEC results indicating the Mw (A) and Mz (B) for the f-PP as a function of peroxide concentration in the feed: (●) Luperox PMA; (□) Lupersol 512; (■) Luperco 212-P75.

Luperox TA-PMA) shows the lowest melt index. In comparing the Luperox pair to Lupersol 512 and Luperco 212, it is clear that the chemical nature of the $R_2\bullet$ radical is an important factor in affecting the melt index. As noted above, the unsaturated group of the R_2 species apparently contributes to cross-linking. In fact, the melt indices for the f-PP samples extruded with Luperox TA-PMA are all lower than for the pure PP sample. This behavior provides strong support for potential cross-linking reactions. While Luperox PMA renders more efficient grafting and is also capable of yielding cross-links, the more reactive $R_1\bullet$ radicals induce significant chain-scission which apparently counteracts the effects of cross-linking.

260

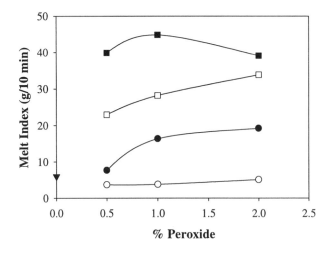

Figure 6. Melt Index of the *f*-PP as a function of peroxide concentration in the feed: (■) Luperco 212-P75; (□) Lupersol 512; (●) Luperox PMA; (○) Luperox TA-PMA. The melt index of pure PP (▼) is shown for comparison.
(Reproduced with permission from reference 25. Copyright 1997 Society of Plastics Engineers.)

Effect of *f*-PP on PP/PA-6,6 Blend Properties. The effect of *f*-PP on the phase dimensions of the blends is illustrated in Figure 7. Figure 7A is a micrograph of the blend containing only pure PP and PA-6,6. The immiscibility of the two polymers results in a phase separated system where the minor phase (PA-6,6) is dispersed as spherical domains a few microns in diameter. Incorporation of 10% *f*-PP into the blend (Figure 7B) results in a noticeable reduction in the minor phase dimensions. Moreover, a 30% addition of *f*-PP (Figure 7C) yields a homogenous surface where the minor phase component is not clearly distinguishable. This "emulsification" of the PA-6,6 into the PP matrix may be attributed to the specific interactions between the carboxyl groups on the *f*-PP and the polar components of the PA-6,6. For example, the carboxyl moieties may interact in a manner similar to that of MAH functionalized PP; chemical reaction with the amine end groups of the PA-6,6 yields PP-b-PA chains(*31*). In addition to block copolymer formation, the carboxyl grafts of the *f*-PP may exhibit specific interactions (e.g., hydrogen bonding) with the polar moieties of the PA-6,6. The presence of chemical and/or specific interactions at the polymer-polymer interface acts to decrease the interfacial tension, resulting in a highly dispersed minor phase.(*20,32*)

Figures 8 and 9 show the respective cooling and subsequent heating DSC thermograms for all of the blend compositions. For the cooling trace of the control blend (curve **A**), two distinct exotherms are apparent at 113°C and 220°C. These exothermic peaks are attributed to the crystallization of the PP and the PA-6,6, respectively. Upon heating, the control blend shows endotherms at 160°C and 255°C, which are attributed to the melting temperatures of PP and PA-6,6, respectively.

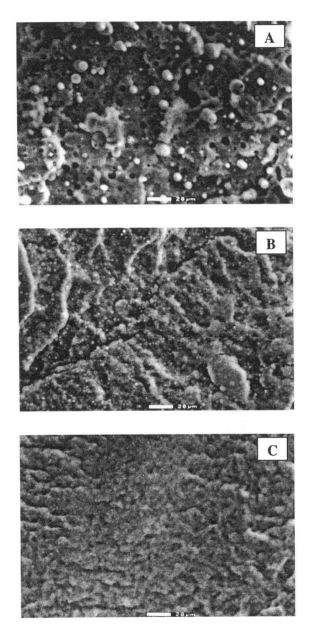

Figure 7. SEM micrographs of PP/PA-6,6 blends containing: (A) 0%, (B) 10%, and (C) 30% *f*-PP.

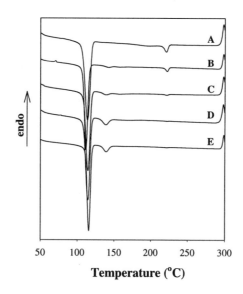

Figure 8. DSC cooling thermograms for the PP/*f*-PP/PA-6,6 blends: (A) 80/0/20, (B) 75/5/20, (C) 70/10/20, (D) 60/20/20, and (E) 50/30/20. (Reproduced with permission from reference 25. Copyright 1997 Society of Plastics Engineers.)

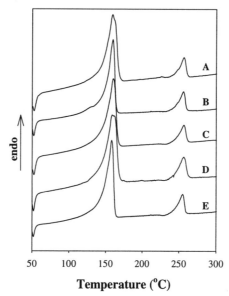

Figure 9. DSC heating thermograms for the PP/*f*-PP/PA-6,6 blends: (A) 80/0/20, (B) 75/5/20, (C) 70/10/20, (D) 60/20/20, and (E) 50/30/20. (Reproduced with permission from reference 25. Copyright 1997 Society of Plastics Engineers.)

Incorporations of 5% and 10% f-PP into the blend result in cooling curves **B** and **C** that exhibit two crystallization peaks for PA-6,6. The high temperature peak is attributed to the original T_c of PA-6,6 at 220°C, while the low temperature peak is a new T_c at 138°C. When 20% and 30% f-PP is incorporated into the blend (curves **D** and **E**), the T_c at 138°C becomes the dominant exothermic event for PA-6,6 crystallization. It is interesting to note that while an 80°C shift in the PA-6,6 T_c is observed, there appears to be no shift in the melting endotherms (Figure 9). This type of behavior may be attributed to a change in the crystallization nucleation mechanism for PA-6,6.(33-36) For example, PA-6,6 crystallization is generally nucleated by a heterogeneous mechanism. However, as the PA-6,6 minor phase becomes finely dispersed into the PP matrix, the number of PA-6,6 phases becomes greater than the number of heterogeneous nucleation sites. This occurrence causes the nucleation of PA-6,6 crystallization to predominantly occur by a homogeneous mechanism, at a lower temperature. To experimentally verify this explanation, a nucleating agent ($CaCl_2$) was incorporated into the minor phase of the blend containing 30% f-PP in an attempt to promote heterogeneous nucleation. DSC thermograms of this blend showed the reemergence of a dominant T_c at 220°C, corresponding to formation of heterogeneously nucleated PA-6,6 crystallites.

The tensile properties of the blends are shown in Figure 10. Although there are no significant enhancements in the initial moduli of the blends with the additions of f-PP, a noticeable increase in the yield stress can be observed. Blends containing 5% and 10% f-PP rupture near the same strain level as the control blend; however, a slight enhancement in the elongation at break is observed with a 20% composition of f-PP. Moreover, blends containing 30% f-PP demonstrate enhanced elongation properties, resembling that of the pure blend components. The changes in the tensile properties of the blends with additions of f-PP may be attributed to a decrease in

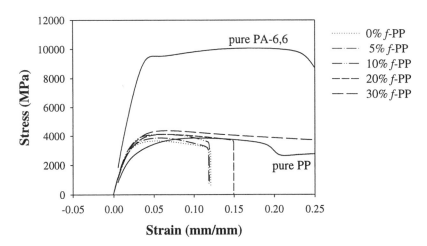

Figure 10. Tensile properties for the PP/f-PP/PA-6,6 blends as a function of f-PP content.

minor phase (PA-6,6) dimensions, increased surface area, and a more strengthened interfacial region due to the chemical and/or specific interactions of the carboxyl groups on the f-PP with the polar moieties of the PA-6,6.

As expected, the impact strength (Figure 11) of the control blend is near the strength of the matrix polymer. However, blends that contain 30% f-PP possess impact properties that approach that of the PA-6,6 minor phase. This behavior suggests that the presence of f-PP in the blends significantly increases the interfacial adhesion (via block copolymer formation) between the matrix and the minor phases. Thus, the compatibilized blends demonstrate properties approaching that of PA-6,6.

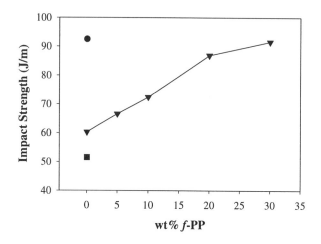

Figure 11. Notched Izod impact results for the PP/f-PP/PA-6,6 blends as a function of f-PP content: (●) PA-6,6; (□) PP, and (▼) 80/20 blend.
(Reproduced with permission from reference 25. Copyright 1997 Society of Plastics Engineers.)

Conclusions

Polypropylene can be functionalized to contain grafted carboxylic acid groups via a reactive extrusion process using functional peroxides. Moreover, the specific grafting process can be tailored and optimized by systematically altering the molecular structure of the peroxides. By affecting the reactivities of the radicals generated from the peroxides, the grafting efficiency of the carboxylic acid group onto the polypropylene backbone, and the polymer degradation process may be controlled.

Based on bond dissociation energy considerations, the relative reactivities of the free radicals fall in the order: methyl ≈ functional alkenyl > ethyl ≈ functional alkyl. The peroxides which yield reactive methyl radicals are more efficient at producing polymeric radicals, relative to the peroxides which generate less reactive ethyl radicals. An increase in the grafting efficiency and PP degradation is apparently attributable to this increase in the production of polymeric radicals. With respect to the functional radicals, the peroxides which yield radicals bearing double bonds have a higher grafting efficiency. This behavior may be partially attributed to an increase in the reactivity of the alkenyl radicals, relative to the alkyl radicals. In addition, the increased grafting efficiency may be further attributed to the potential for the polymeric radicals to add across the double bond of the alkenyl groups. Therefore, this study clearly demonstrates that subtle changes in the molecular structure of peroxides can yield a range of radical reactivities that may be exploited in the precise control of chemical reactions within a twin-screw extruder.

Moreover, the incorporation of the f-PP into PP/PA-6,6 (80/20) blends results in an efficient emulsification of the minor phase as shown by SEM. This enhanced minor phase dispersion significantly alters the crystallization process of the PA-6,6 by lowering the T_c approximately 80°C. The presence of the f-PP at the interfacial region is also demonstrated by the significant improvement in the impact and tensile properties of the blends.

Acknowledgments

Support for this work was provided by Elf Atochem North America, Inc. We gratefully acknowledge Professor A. Rudin for the high temperature GPC results.

References

1)Brown, S. B.; Orlando, C. M. *Encyc. of Polym. Sci. Eng.*; 2nd ed., 1988; Vol. 14.
2)Xanthos, M. *Reactive Extrusion: Principles and Practice*; Hanser Publishers: New York, 1992.
3)Tucker, C. S.; Nichols, R. J. *Plastics Engineering* **1987**, 27.
4)Stuber, N. P.; M., T. *Polym. Process Eng.* **1985**, *3*, 71.
5)Matsuoka, N.; Matsumoto, H.; Hori, Y.; Sano, K.; Ijichi, I. *US 4,487,897*, Nitto Electric Industrial, **1984**.
6)Tang, Y.; Tzoganakis, C.; Hamielec, A. E.; Vlachopoulos, J. *Adv. Polym. Tech.* **1989**, *9(3)*, 217.
7)Suwanda, D.; Lew, R.; Balke, S. T. *J. Appl. Polym. Sci.* **1988**, *35*, 1019-1032.
8)Tzoganakis, C.; Vlachopoulos, J.; Hamielec, A. E. *Polym. Eng. Sci.* **1988**, *28*, 170-180.
9)Suwanda, D.; Lew, R.; Balke, S. T. *J. Appl. Polym. Sci.* **1988**, *35*, 1033-1048.
10)Cartasegna, S. *Rubber Chem. Technol.* **1986**, *49*, 722.
11)Greco, R.; Maglio, G.; Musto, P. *J. Appl. Polym. Sci.* **1987**, *33*, 2513.
12)Gaylord, N. G.; Mishra, M. K. *J. Polym. Sci. Polym. Lett. Ed.* **1983**, *21*, 23.
13)Gaylord, N. G.; Mehta, R. *J. Polym. Sci. Polym. Chem. Ed.* **1987**, *26*, 1189.
14)Ho, M.; Su, A. C.; Wu, C. H. *Polym. Prepr. (Am. Chem. Soc., Div. Polym. Chem.)* **1992**, *33*, 941.
15)Liu, N. C.; Baker, W. E.; Russel, K. E. *J. Appl. Polym. Sci.* **1990**, *41*, 2285-2300.
16)Greco, R.; Maglio, G.; Musto, P.; Scarinzi, G. *J. Appl. Polym. Sci.* **1989**, *37*, 777.
17)Simmons, A.; Baker, W. E. *Polym. Eng. Sci.* **1989**, *29*, 1117.
18)Bratawidjaja, A. S.; Gitopakmoyo, I.; Watanabe, Y. *J. Appl. Polym. Sci.* **1989**, *37*, 1141.

19)Minoura, Y.; Ueda, M.; Mizunuma, S.; Oba, M. *J. Appl. Polym. Sci.* **1969**, *13*, 1625.

20)Ide, F.; Hasegawa, A. *J. Appl. Polym. Sci.* **1974**, *18*, 963-974.

21)Lin, J. S.; Sheu, E. Y.; Jois, Y. H. R. *J. Appl. Polym. Sci.* **1995**, *55*, 655-666.

22)Heinen, W.; Rosenmoller, C. H.; Wenzel, C. B.; de Groot, H. J. M.; Lugtenberg, J.; van Duin, M. *Macromolecules* **1996**, *29*, 1151.

23)Russell, K. E.; Kelusky, E. C. *J. Polym. Sci. Polym. Chem. Ed.* **1988**, *26*, 2273.

24)Rengarajan, R.; Parameswaran, V. R.; Lee, S.; Vicic, M.; Renaldi, P. L. *Polymer* **1990**, *31*, 1703.

25)Manning, S. C.; Moore, R. B. *J. Vinyl Add. Tech.* **1997**, *3*, 184-189.

26)DeNicola, A. J.; Wei-Berk, C. H.; Hogt, A. H.; Jelenic, J.; Meijer, J. *US 5,447,985*, Akzo Nobel N.V., **1995**.

27)*Polymer Handbook*; Brandrup, J.; Immergut, E. H. Eds.; John Wiley & Sons: New York, NY, 1975; 2nd ed.

28)Kirchgessner, R. J.; Kamath, V. R.; Sheppard, C. S.; Stromberg, S. E. *Mod. Plast.* **1984**, *61*, 66.

29)Dorn, M. *Adv. Polym. Technol.* **1985**, *5*, 87.

30)Hogt, A. H. *2nd International Conference on Advances in Additives and Modifiers for Polymer Blends*, Philadelphia, PA, **1993**.

31)Grulke, E. A. *Polymer Handbook*; Brandrup, J. and Immergut, E. H. eds.; Wiley Interscience: New York, NY, 1989, pp 519-559.

32)Legras, R.; Marechal, P.; Coppens, G.; Dekoninck, J. *J. Polym. Sci. Part A: Polym. Chem.* **1995**, *33*, 757-766.

33)Hosoda, S.; Kojima, K.; Kanda, Y.; Aoyagi, M. *Polym. Networks Blends* **1991**, *1*, 51.

34)Morales, R. A.; Arnal, M. L.; Muller, A. J. *Polym. Bull.* **1995**, *35*, 379.

35)Santana, O. O.; Muller, A. J. *Polym. Bull.* **1994**, *32*, 471.

36)Matos, M. E.; Rosales, C.; Muller, A. J.; Favis, B. D. *ANTEC* **1997**, 2552-2556.

37)Holsti-Miettinen, R.; Seppala, J.; Ikkala, O. T. *Polym. Eng. Sci.* **1992**, *32*, 868-877.

Chapter 17

Layer-by-Layer Polycondensation of Nylon-66 with Normal Molecular Orientation by Alternating Vapor Deposition Polymerization

Huili Shao and Norimasa Okui

Department of Organic and Polymeric Materials, Tokyo Institute of Technology, Ookayama, Meguroku, Tokyo 152, Japan

Nylon-66 thin films were prepared by alternating vapor deposition polymerization(AVDP) method by the layer-by-layer polycondensation. In the AVDP process, two kinds of bi-functional monomers, such as Adipyl chloride monomer (AC) and Hexamethylene diamine monomer (HMDA) are deposited alternatively onto a substrate. At first, AC is deposited onto the amino covered substrate. AC can react with the amino groups on the substrate and the substrate surface is covered with the chain ends of AC. Next, HMDA is deposited onto the mono-molecular layer of the AC yielding the film surface to be covered with the chain ends of HMDA. The resultant nylon films showed the highly ordered structure with the molecular orientation tilting to the substrate. The thickness and the molecular weight for the nylon film increased linearly with the number of deposition reaction cycles. The molecular weight of the film is proportional to the film thickness.

Vapor deposition polymerization (VDP) is a useful and promising method for fabrication of functional polymeric thin films with thermal stability such as polyimide thin films [1,2]. VDP can also be used to fabricate polyurea films having piezoelectricity and pyroelectricity [3]. Compared with the Langmuir-Blodgett (LB) method and other preparation methods of polymeric thin films, VDP has a great advantage since the method is a simple dry process [1]. In addition, molecular orientation in a VDP film can be controlled. For example, the VDP method can produce polyamide [4-7] and polyimide [8,9] films with high molecular orientation perpendicular to the substrate surface.

The present paper deals with a basic idea of the alternating VDP (AVDP), in which, two kinds of bifunctional monomers, such as a dicarboxylic chloride monomer and a diamine monomer, which can react easily at their chain ends, are deposited alternatively onto a substrate. The mechanism of adsorption/desorption of monomers onto the substrate is studied and the feasibility of AVDP method for preparation of an ultra-high molecular weight nylon-66 thin film with a layer-by-layer structure is investigated.

Experimental

Hexamethylene diamine (HMDA, Tm=42°C) and Adipyl chloride (AC, liquid state at room temperature and atmospheric pressure) were provided by Tokyo Kasei Co. Ltd. and used without further purification. P-type silicon wafers with a (100) plane (Mitsubishi Materials Silicon Co.) were used as substrates. Before the AVDP process, the substrates were treated by a silane-coupling agent (γ -aminoethyl-aminopropyl-trimethoxy silane) to form an activated sublayer surface covered with the amino end group [7], which could react with the chain ends of the dicarboxylic chloride monomer.

Figure 1 shows the scheme of the experimental apparatus. First, the system was evacuated to remove air by means of opening the valve C for a certain period of time. AC monomer kept at 65°C was introduced into the system by closing valve C and opening valve A for a certain period of time in order to deposit the monomer onto the activated substrate. Next, valve A was closed and valve C was opened to evacuate the residual monomers from the system for a certain period of time. After closing valve C, valve B was opened to introduce the vapor of HMDA monomer kept at 55°C for a certain period of time in order to deposit the monomer onto the substrate. Valve C was opened again after closing valve B in order to eliminate the excess monomers and by-products from the system. This reaction cycle was repeated to get a film with the desired thickness.

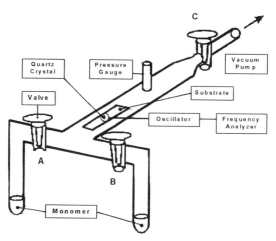

Figure 1   Schematic illustration of the experimental apparatus for AVDP process

The adsorption/desorption behaviors of monomers onto the substrate were investigated by monitoring the change in the amount of adsorbate on the surface of Au-coated quartz crystal substrate (QCM). The substrate surface had been covered with the amino end groups of the silane-coupling agent by a surface treatment process prior to the deposition of monomers.

IR spectra were obtained with transmission and RAS methods using a Fourier transform IR system (Jasco FT-IR 7000) at room temperature in a dry nitrogen atmosphere. X-ray diffraction patterns were recorded in a θ-2θ reflection mode by a Rigaku RAD-B diffractometer using Ni filtered CuKα X-rays. The film thickness was measured by an ellipsometer (EL-8, Optec Co. Ltd.) with an incident angle of 70° of the light from a He-Ne laser (632.8nm). The films on the substrate were scraped off from the substrates for determination of the molecular weight by viscometry in 90% formic acid at 25°C [10]. Differential scanning calorimetric (DSC) measurements were carried out using a Shimazu DSC-50 equipment in the nitrogen atmosphere at a heating rate of 10°C min^{-1}.

Results and Discussion

Deposition behavior of AC monomer. Figure 2 shows the change in the amount of AC monomer adsorbed on the amino-covered quartz crystal substrate kept at 70°C as a function of the time for monomer supply and for evacuation after stopping the monomer supply. When AC monomer is introduced into the system for 10 seconds, the amount of the adsorbed AC monomer on the substrate increases to the stage a in figure 2. Then, the monomer supply is stopped and the residual monomer and by-product HCl are evacuated from the system for 1 minute. During the above process, the amount of adsorbate remains almost constant, indicating that no admolecules are re-evaporated from the substrate. In other words, AC molecules are adsorbed on the substrate by chemical bonding (chemisorption), which are attributed to the condensation reaction between the carboxylic acid chloride groups in AC and the amino groups on the substrate surface. When AC monomer is introduced into the system for another 10 seconds, the amount of the adsorbed AC monomer is slightly increased and is saturated to the stage b. Then, it keeps at the constant value even if the subsequent evacuation and the monomer supply are proceeded (stage c to stage d). The saturated value 1.0×10^{-6}kg/m^2 at b is coincident with the areal density of AC monomolecular layer with the normal orientation to the substrate. These results indicate that during the first 10s deposition, the amount of the AC monomer is not enough to form the complete monomolecular layer of AC monomer as shown in figure 3. However, during the next 10s deposition, the non-adsorbed sites on the substrate are occupied by the AC monomer yielding the complete monomolecular layer.

Deposition behavior of HMDA monomer. Figure 4 shows the change in the amount of HMDA monomer adsorbed on the quartz crystal surface, which has been covered previously with the AC monomolecular layer, as a function of the time for monomer supply and evacuation after stopping the monomer supply. When HMDA is introduced into the system for 10s, the amount of adsorbate of HMDA monomer on the substrate surface increases rapidly and tends to the stage a in figure 4. After

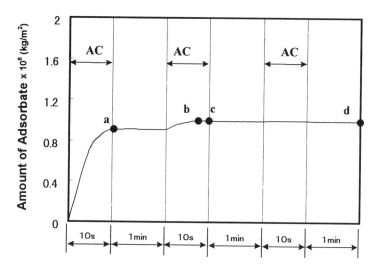

Figure 2. Change in the amount of adsorbate for AC on the quartz crystal at 70°C as a function of the monomer supplying time (10s) and the evacuation time (1min) after stopping the monomer supply.

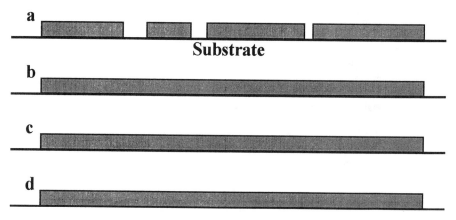

Figure 3. Schematic illustration of adsorption behavior of AC monomer based on the results in figure 2.

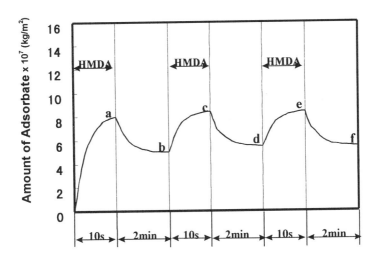

Figure 4. Change in the amount of admolecules of HMDA on the quartz crystal with AC monomolecular layer surface at 70°C as a function of the monomer supplying time (10s) and the evacuation time (2 min) after stopping the monomer supply.

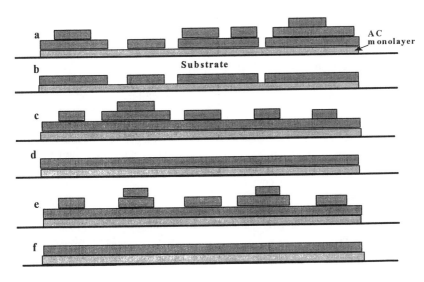

Figure 5. Schematic illustration of the adsorption/desorption mechanism for HMDA based on the results in figure 4.

stopping the monomer supply and evacuating system for 2 minutes, the amount of adsorbate decreases to the stage b. It can be depicted that the adsorbate at the stage a is attributed to both chemisorption and physisorption of HMDA and the physisorbed molecules are re-evaporated entirely from the substrate at the stage b as shown in figure 5. The adsorbed amount at the stage b is attributed to the chemisorption of HMDA. When HMDA is successively introduced into the system for the second 10s, the increase in the amount of adsorbate is due to not only physisorption but also chemisorption of HMDA (stage c). After terminating the monomer supply, the physisorbed HMDA monomers are evacuated again from the system for 2 minutes. The amount of adsorbate at the stage d is slightly larger than that at the stage b. The increment of the HMDA adsorbed on the substrate by the subsequent monomer supply (stage d, e) is equal to the decrement of the monomer which is due to the desorption of the monomer from the substrate after stopping the monomer supply (stage e, f). Consequently, the amount of adsorbate is decreased to a certain value of $5.5 \times 10^{-7} kg/m^2$ (stage d and f), which is coincident with the areal density of the monomolecular layer of HMDA with a normal molecular orientation on the substrate. These results indicate that the first adsorption layer of HMDA monomer onto the substrate surface consists of the chemisorption of HMDA based on the reaction between one of the chain ends of HMDA monomer and the chain ends of AC monomer on the substrate surface with eliminating HCl. The excess HMDA monomers are physisorbed onto the first HMDA chemisorbed layer. However, the complete chemisorbed monomolecular layer is achieved by the couple of monomer supply and the evacuation of the excess physisorbed monomer from the chemisorbed layer.

Alternating deposition process of AC and HMDA. According to the above results of deposition behaviors for AC and HMDA monomers, the essential deposition times of monomers for forming complete monomolecular layers can be estimated experimentally, such as 20s monomer supply and 5 min evacuation in the system after stopping the monomer supply. Figure 6 shows the change in the amount of admolecules on the quartz crystal substrate covered with the amino groups of the silane coupling agent during the alternating vapor deposition of AC and HMDA. When AC monomer is introduced into the system for 20s, the adsorbed amount is reached to a saturated value of about 1.0×10^{-6} kg/m^2 (stage a), which is the identical with that at the stage b in figure 2. As discussed above, the adsorption of AC is a chemisorption process. After the termination of the AC monomer supply, the adsorbed amount remains the constant value even if the system is evacuated for 5 minutes (stage b). These results indicate that a complete AC monomolecular layer is formed on the amino-covered surface. Next, HMDA is introduced into the system and is adsorbed on the AC monomolecular layer. The adsorption increases to stage c based on the chemisorption and physisorption of HMDA. After terminating the HMDA supply, the increment of the HMDA adsorbate is about 5.5×10^{-7} kg/m^2 (stage d), which is equal to that at the stage f in figure 4. The total amount of adsorbate at the stage d coincides with the areal density of nylon-66 crystal density with the normal orientation.

Figure 7 shows the film thickness for the nylon 66-film measured by QCM in situ during AVDP as a function of the number of reaction cycles (i.e. the number of bilayers). The film thickness increases almost linearly with the number of alternating

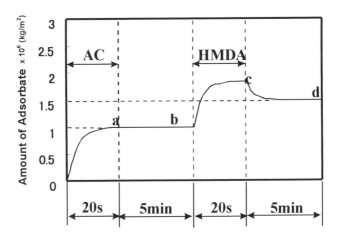

Figure 6. Change in the amount of admolecules on the quartz crystal substrate covered with amino groups during the alternating vapor deposition of AC and HMDA. The substrate temperature is kept at 70°C.

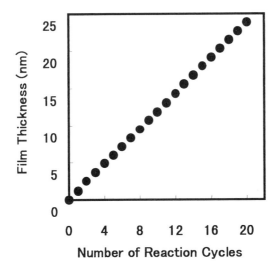

Figure 7. Film thickness of nylon-66 on the quartz crystal monitor (in situ measurement) as a function of the number of deposition reaction cycles. The thickness was estimated from the amount of adsorbate and the density of nylon-66 crystal.

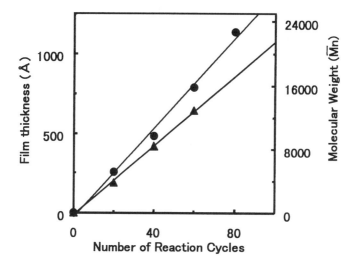

Figure 8 Film thickness (●) and molecular weight (▲) of nylon 66 films prepared at 60°C by the AVDP method as a function of the number of reaction cycles.

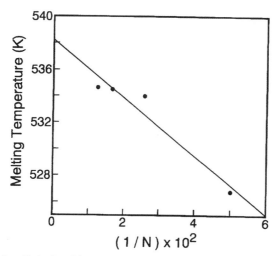

Figure 9. Relationship between the melting temperature of the AVDP film and the reciprocal of the number of reaction cycles.

deposition reaction cycles (i.e. the degree of polymerization). From the slope in figure 7, the layer thickness is 1.21nm which value is associated with the repeating unit length of nylon-66. The film thickness (off-line measurement by an ellipsometry) and molecular weight (determined by viscometry) are shown in Figure 8 as a function of the number of reaction cycles. The film thickness and the molecular weight increase linearly with the number of reaction cycles. That is, the molecular weight of nylon-66 film is proportional to the film thickness. From the slope in figure 8, the layer thickness is 1.31nm which value is associated with the repeating unit length of nylon-66. These results clearly indicate that the increase in the thickness of the film is due to a chain-extending process from the substrate surface without a chain folding. In other words, the film is constructed with extended chains aligned normal to the substrate. This is also confirmed by DSC data. DSC thermogram of AVDP film showed a single melting peak corresponding to the melting point of the nylon-66 crystal. Figure 9 shows the melting point as a function of the reciprocal of the number of reaction cycle (degree of polymerization). The melting point shows a linear relationship with the reciprocal of the degree of polymerization, indicating the melting behavior of the extended chain crystal.

Mechanism of alternating deposition of AC and HMDA. According to the above adsorption/desorption behavior, we can figure out the mechanism of the alternating vapor deposition polymerization (AVDP) as shown in figure 10. As described previously, before the AVDP process, it is necessary to prepare an activated sublayer with the amino end groups on the substrate surface (Figure 10 (a)). And each reaction cycle of alternating deposition polymerization consists of a two-step procedure. In the first step, the evaporated AC monomer is introduced to the system. The molecule may adsorb randomly and travel on the substrate surface. At the stage of molecular deposition onto the substrate, molecular orientation is strongly dependent on the substrate temperature [1,11]. When the substrate temperature is kept at a sufficiently high (such condition adopted in this paper), most molecules tend to the normal orientation. The probability of reaction at the both chain ends at the same time on the substrate surface may be negligible. When one of the chain ends reacts on the substrate, the monomer molecule sticking with one of its chain end (hair molecule) on the substrate tends to form upright molecular orientation. Such upright molecules can not bend sharply to the substrate surface because of restriction of their bonding angle and conformational energy. Consequently, one of its chain ends can react with the amino groups on the surface with the elimination of HCl (Figure 10 (b)). The chemisorbed molecules exhibit higher molecular orientation perpendicular or tilting to the substrate. The by-product and the excess AC monomer are evacuated from the system. The AC monolayer is formed on the activated sublayer and its surface is covered with the chain ends of AC (Figure 10 (c)).

In the second step, the HMDA is introduced to react with the chain ends of AC to complete the first reaction cycle (Figure 10 (d)) and simultaneously the excess diamine and HCl are removed from the system by vacuum-pumping. This results in the amido linkage (condensation reaction) yielding the activated surface covered with the amino groups that will be valid to initiate another reaction cycle (Figure 10 (e)), and so forth, until the desired number of bilayers are constructed. Accordingly, the repetition of

276

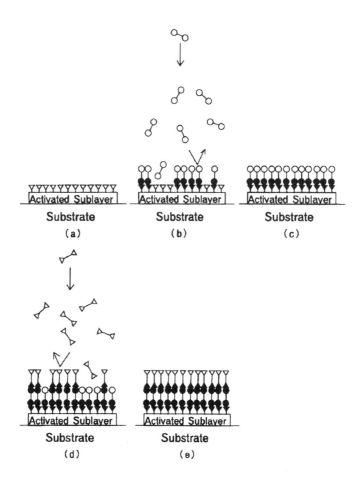

Figure 10. Illustration of the mechanism of alternating vapors deposition polymerization of nylon system. (a): An activated sublayer on the substrate. Open triangles refer to the amino end groups of the silane-coupling agent. (b): The AC monomer reacts with the amino groups on the sublayer surface. Open circles refer to the chain ends of dichloride monomer. Solid cycles and triangles are the reacted functional groups. (c): AC monolayer covered with the chain ends of AC. (d): The HMDA monomer reacts with the chain ends of AC. Open triangles refer to the diamine chain ends. (e): The first reaction cycle is completed to form an activated surface covered with the amino groups that can initiate another reaction cycle.

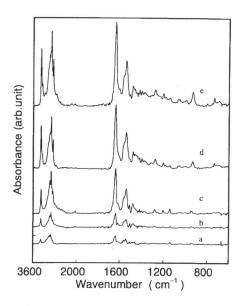

Figure 11. Transmission infrared spectra of nylon-66 thin films prepared by the alternating vapor deposition polymerization at a substrate temperature of 70°C and at different number of reaction cycles (N): (a) N=10; (b) N=20; (c) N=40; (d) N=60; (e) N=80

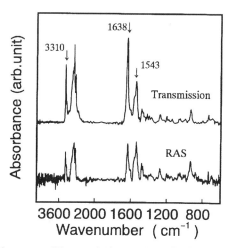

Figure 12. Infrared spectra (Transmission and RAS) for nylon-66 thin films prepared by the alternating vapor deposition polymerization for 80 reaction cycles at a substrate temperature of 70°C.

alternating reaction can give rise to an increase in the molecular weight of the polymeric film without any chain folding because of the repetition of the monomolecular layer-by-layer epitaxial process.

Structure and molecular orientation of nylon-66 film. Figure 11 shows the transmission IR spectra for the film prepared at a substrate temperature of 70°C by the alternating vapor deposition polymerization of AC and HMDA at different number of reaction cycles. The AVDP films show characteristic amide peaks at 3310 cm^{-1} (N-H stretching vibration), 1638 cm^{-1} (C=O stretching vibration) and 1543 cm^{-1} (N-H bending vibration). No peaks for the residual monomer are found. These characteristic amide peaks increase linearly with an increase in the reaction cycles. The peak at 1638cm^{-1} due to C=O stretching vibration ($v_{C=O}$) having a transition moment perpendicular to the molecular chain axis appears strongly in the transmission spectrum but weakly in the RAS spectrum as shown in figure 12. On the other hand, the peak at 1543cm^{-1} due to N-H bending vibration (δ_{N-H}) having a transition moment parallel to the molecular chain axis appears to be in a contrary result. Figure 13 shows the intensity ratio of the peaks at 1638cm^{-1} and at 1543cm^{-1} in the IR transmission spectra as a function of the number of reaction cycle. The intensity ratio keeps almost the constant value, indicating that the molecular orientation in the film does not change during the layer-by-layer polycondensation by AVDP method.

Figure 14 shows the wide angle X-ray diffraction (WAXD) pattern of nylon-66 film prepared by the AVDP method. Four sharp diffraction peaks ranging from small to wide angle can be associated with the reflections of the layered structure from first order to fourth order, indicating that the polymeric film consists of a highly ordered layered structure. The peak at a diffraction angle of about 8.7° corresponds to a spacing of 1.01nm. The spacing is shorter than that calculated from the fully extended chain (1.72nm) [12] but roughly corresponds to the bilayer thickness estimated from the slope in figure 7 by QCM (1.21nm) and the slope in figure 8 by ellipsometry (1.31nm). These results indicate that the molecular chains in nylon-66 AVDP film are tilted at about 54° to the substrate normal. The inclination of the molecules could be associated with the hydrogen bonding between the amide groups of the adjacent molecules.

The film surface showed very flat in the order of optical microscopic scale and the domain structure was found in the transmission micrograph by surface replica method. The surface of the domain was very flat and its size was about 600-3000Å. These results indicate that a uniform nylon-66 thin film with well defined thickness and layer-by-layer structure is formed by the alternating deposition of AC monomer and HMDA monomer.

Application of AVDP method to other polycondensation systems. As discussed above, nylon 66 thin films with a layered structure and a desired molecular weight are prepared using the AVDP method. According to the principle of alternating layer-by-layer polycondensation, we can apply this method to other nylon systems, such as the other binary nylon system (for example nylon-106, nylon-1010 and nylon-79) and a quadruple nylon system (nylon-66610, nylon-661010 and nylon-7999) [13]. All these films prepared by the AVDP method showed a layer-by-layer structure similar to the nylon-66 system.

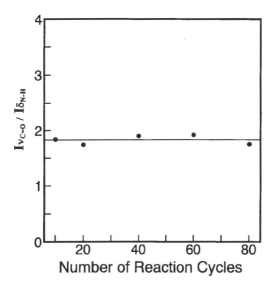

Figure 13. Absorption intensity ratio $I_{\nu_{C=O}}$ / $I_{\delta_{N-H}}$ in the IR transmission spectra for nylon 66 films prepared by the AVDP method as a function of the number of reaction cycles

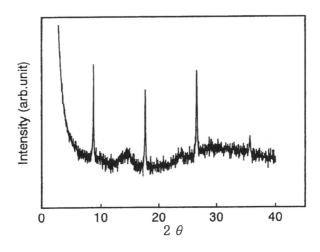

Figure 14. WAXD pattern of nylon-66 thin film deposited onto the substrate at 70°C for the number of reaction cycles of N=20

280

Conclusion

The alternating vapor deposition polymerization is a useful method to produce nylon-66 thin film with a layered structure having a higher molecular orientation tilted to the substrate surface. The film formation processes involve molecular adsorption and desorption due to chemisorption and physisorption. The dicarboxylic chloride (AC) monomers are chemisorbed on the amino covered substrate to form a monomolecular layer with a higher molecular orientation and no physisorption of AC is found. On the other hand, the diamine (HMDA) monomers are not only chemisorbed on the AC layer but also physisorbed on its chemisorbed monolayer. However the complete chemisorbed monomolecular layer of HMDA can be formed by repeating the adsorption and desorption process of HMDA. The repetition of alternating deposition reaction of two monomers can give rise to an increase in the molecular weight of the film without any chain folding. The thickness and molecular weight of polymeric thin film can be effectively controlled by the number of reaction cycles during the alternating vapor deposition polymerization process. In addition, the AVDP method can be applied to prepare a multi-nylon system such as nylon-66610 and nylon-661010 and an ultra-high molecular weight polymer in a polycondensation system.

References

1.   Kubono, A.; Okui, N., Prog. Polym. Sci., 1994, 19, 389
2.   Takahashi, Y.; Iijima, M.; Oishi, Y.; Kakimoto, M.; Imai, Y., Macromolecules, 1991, 24, 3543
3.   Fukada, E.; Wang, X.S.; Hattori, T.; Iijima, M.; Takahashi, Y, Ferroelectrics, 1994, 151, 281
4.   Kubono, A.; Okui, N.; Tanaka, K.; Umemoto, S.;.Sakai, T., Thin Solid Films, 1991, 199, 385
5.   Kubono, A.; Kanae, H.; Umemoto, S.; Sakai, T.; Okui, N., Thin Solid Films, 1992, 215, 94
6.   Kubono, A.; Yuasa, N.; Shao, H.L.; Umemoto, S.; Okui, N., Thin Solid Films, 1996, 289, 107
7.   Shao, H.L.; Umemoto, S. Kikutani, T.;.Okui, N., Polymer, 1997 38, 459
8.   Kubono, A.; Higuchi, H.; Umemoto, S.; Okui, N.,Thin Solid Films, 1993, 229, 133
9   Kubono, A.; Higuchi, H.; Umemoto, S.; Okui, N.,Thin Solid Films, 1993, 232, 256
10.   Tayer, G.B., J.Am.Chem.Soc., 1947, 69, 635
11.   Tanaka, K.; Okui, N.; Sakai, T., Thin Solid Films, 1991, 196, 137
12.   Bunn, G.W.;Garner, E.V.; Proc. Roy. Soc. (London),   1947, A189, 39
13.   Nagai. A.; Shao, H.L.; Umemoto, S.; Kikutani, T.; Okui, N.; Polymer Preprints Jpn., 1997, 46, 3369

INDEXES

Author Index

Subject Index

A

283